CHOOSING AND USING FARM MACHINES

BRIAN WITNEY

CHOOSING
AND USING
FARM MACHINES

Longman Scientific & Technical

Copublished in the United States with
John Wiley & Sons, Inc., New York

Longman Scientific & Technical,
Longman Group UK Limited,
Longman House, Burnt Mill, Harlow,
Essex CM20 2JE, England
and Associated Companies throughout the world.

Copublished in the United States with
John Wiley & Sons, Inc., 605 Third Avenue, New York, NY 10158

First published 1988

British Library Cataloguing in Publication Data
Witney, Brian
Choosing and using farm machines.
1. Agricultural machinery
I. Title
631.3   S675

ISBN 0-582-45600-2

Library of Congress Cataloging-in-Publication Data
Witney, Brian, 1938–
Choosing and using farm machines/Brian Witney. p.   cm.
Bibliography: p. cm.
Includes index.
ISBN 0-470-21028-1 (Wiley, USA only).
1. Agricultural machinery. I. Title.
S675.W63 1988
631.3 – dc19 87-37485 CIP

Set in Linotron 202 10/12 pt Times
Produced by Longman Singapore Publishers (Pte) Ltd.
Printed in Singapore

CONTENTS

ACKNOWLEDGEMENTS

We are grateful to the following for permission to reproduce copyright material:

AFRC Engineering for figs. 3.2 (Nation 1978) and 6.10 (Brown & Charlick 1972); American Society of Agricultural Engineers for figs. 2.19 (Dupuis 1959), 2.20 (Purcell 1980, adapted), 6.5 (Pierce 1960), 7.6, 7.10 (Brixius & Wismer 1978), 7.20 (Zoz 1972), 7.21 (Zoz 1974) & 8.20 (Shoup 1983) and tables 2.10 (Dupuis 1959), 4.4, 4.11 (ASAE 1982) & 8.5 (Shoup 1983); ASHRAE Inc. for fig. 2.27 (ASHRAE 1985); Blackwell Scientific Publications Ltd. for figs. 5.6 & 5.7 (Jenkins & Storey 1975); Blackwell Scientific Publications Ltd. and the author, C. Culpin (and V. Baker) for fig. 2.9 and table 4.3 (Culpin 1975); British Standards Institution for fig. 2.24 (ISO 2631); BTR Industries for figs. 7.9 & 7.13 (Inns & Kilgour 1978) (c) Dunlop Ltd.; the author, J. Cathie for figs. 3.15, 3.16 & 3.17 and tables 3.4 & 3.5 (Sturrock *et al* 1977); Centre for Farm Management for fig. 9.9 (Bright 1986) and tables 9.10, 9.11 & 9.12 (Crabtree 1984); Department of Agricultural Economics (Wye College) for table 2.1 (Nix 1986); the author, M. J. Dwyer for tables 7.3 (Dwyer, Comely & Evernden 1975), 7.4, 7.5 & 7.6 (Dwyer 1984); East of Scotland College of Agriculture for figs. 2.6 (SIAE), 5.8, 5.9, 5.10 (Elrick 1982), 6.10 (McGechan 1985/2), 6.11a & b (Spencer *et al* 1985) & 9.1 (Anderson 1986) and tables 3.2, 5.6 (Elrick 1982), 6.4 (Eradat Oskoui 1986), 8.1 (Jeffrey 1981) & 9.2 (SAC 1986); the author, E. B. Elbanna for fig. 7.23 (Elbanna 1986); Elsevier Scientific Publishing Co. for fig. 8.5 (Soane *et al* 1980–1981) and table 2.16 (Owen & Hunter 1983); Farm Building Progress for tables 2.4, 2.5 (Cermak & Ross 1977), 2.9 (Cermak & Ross 1980) & 2.13 (Cermak & Ross 1978); Her Majesty's Stationery Office for fig. 2.30 (HSE 1980/1); the author, F. M. Inns for fig. 8.11 (Inns 1985); Institute of Occupational Hygiene for fig. 2.29 (Spencer 1961); Institution of Agricultural Engineers for figs. 1.1 (Arthey 1986), 2.16 (McGechan 1984), 2.21 (Zegers 1985) & 4.8 (Palmer 1984) and tables 1.3, 1.4 (White 1975) & 2.7 (Easterby 1975); Jordbrukstekniska Institutet, Uppsala for fig. 2.26 (Eriksson 1973); the author, D. Kirkpatrick for table 4.5 (Kirkpatrick 1979); the author, J. B. Liljedahl for fig. 9.5 (Liljedahl *et al* 1979); Massey Ferguson for fig. 2.18, copyright 1984; Methuen & Co. Ltd. for fig. 2.15 (Welford 1968); Ministry of Agriculture Fisheries and Food for figs. 2.25 (Rosseger & Rosseger 1960), 5.5 (ADAS 1976), 6.8 (Anon 1982), 6.12 (Philips & O'Callaghan 1974), 6.13 (Smith *et al* 1981), 7.11 (Gee-Clough 1980, adapted), 8.14 (Kofoed 1984) & 9.8 (Barrett 1985) and tables 2.14 (Bird 1978), 4.2 (MAFF), 5.4 (ADAS 1976), 5.5 (McGechan 1985/1), 9.4, 9.5, 9.6 (Barrett 1982) & 9.8 (Barrett 1985); North of Scotland College of Agriculture for fig. 2.28 (Watson 1979); Pergamon Books Ltd. for table 2.2 (Buckett 1981); Pitman Publishing for figs. 2.1 & 2.2 (Currie 1972); Power Farming for figs. 3.4, 3.7, 3.8 (Kepkay 1972), 8.4 (Dwyer 1982), 8.8 (Mertins 1978), 8.16 (Schrader 1982); the author, R. W. Radley for table 2.15 (Monk *et al* 1984); Scottish Farm Buildings Investigation Unit for fig. 2.22; the author, W. T. Singleton for tables 2.6 & 2.8 (Singleton 1974); Smithsonian Institution Press for table 6.2 from *Smithsonian Meteorological Tables* 1985; the author, E. A. Spackman for table 6.5, 6.6 & 6.7 (Spackman 1983); Statistical Office of the European Communities for tables 1.1 & 1.5 (Anon 1986/2); the author, R. M. Stayner for figs. 8.2 (NIAE 1985) & 8.15 (NIAE 1983) and tables 2.12 (Stayner & Bean 1975), 8.2 & 8.3 (NIAE 1985); Tavistock Publications Ltd. for figs. 2.12 & 2.13 (McLeod & Staffurth 1968); VDI-Verlag GmbH for fig 7.14 a, b & c (Sohne 1959); the author B. D. Witney for figs. 7.15, 7.16, 7.17, 7.18 (Terratec 1982), 8.17, 8.18 & 8.19 (Hagger 1986).

AUTHOR'S ACKNOWLEDGEMENTS

The scientific contributions of dedicated researchers in a wide range of disciplines are much appreciated. In bringing together numerous excellent technical papers and bulletins, it is inevitable that some concepts, applications and examples are best expressed by closely following the originator's phraseology. In particular, special mention is due to the following: the late R M Currie on work study; D R Hunt on machinery performance; W Bowers on workshop care of machinery; E Audsley on present annual costs of machinery ownership; F M Zoz on the tractor drawbar performance predictor; M J Dwyer on tractive performance, and on engine limited and slip limited power output; T F Funk on farmer buying behaviour; F M Barrett on arable work scheduling; and J R Crabtree on machinery finance options.

Much of the background research for this book was undertaken as part of a doctoral research programme on machinery management. The author wishes to pay tribute to the involvement of K Eradat Oskoui, E B Elbanna and T Saadoun in the analytical appraisal of soil moisture content and of soil workdays, of soil strength and of crop yield losses through untimely operations, and of integer linear programs for farm operational scheduling, respectively. Appreciation is expressed to my colleagues, P M Wilson for his unstinting efforts in the production of the line drawings, and A Langley for his critical review of the manuscript.

This text is dedicated to my father, a highly respected agricultural economist, and to Andrew Lyon, an enterprising dairy farmer, who jointly influenced my future career; but it is only through the inspiration, encouragement and support of my wife, Maureen, that the blend of theory and practice in this manuscript has reached fruition.

B D W
October, 1987

NOTATION

a, a_{rms}	acceleration; root mean square acceleration, m/s^2
a_1	intercept constant, dim.
A	area in contact, m^2
A	area, ha
A_o, A_t	present annual ownership cost, before and after tax, £
A_{peak}, A_{rms}	peak and root mean square amplitude, m
AR	accumulated repair and maintenance cost, £
b	tyre section width, m
b_f	length of shortest field boundary, m
b_1	slope constant, mm
BC	balancing charge, £
BD	soil aggregate breakdown, %
c	cohesion, kN/m^2
c_1, c_2	soil type constants
C_a, C_r	actual and rated harvester throughput capacity, t/h
C_c	calendar rate of work, ha/day
C_{dwt}	coefficient of dynamic weight transfer, dim.
C_f	first year correction factor, dim.
C_o	overall rate of work (area capacity), ha/h
$(C_{RR})_f$, $(C_{RR})_r$	coefficient of rolling resistance for front and rear wheels, dim.
C_T	coefficient of traction, dim.
$(C_T)_{max}$	maximum coefficient of traction, dim.
CA	capital allowance, £

CF	cash flow, £
CI	cone penetration resistance, MPa
d	decimal rate of depreciation, dim.
d	undeflected tyre diameter, m
d_f	annual depreciation factor, dim.
d_1	distance travelled in work, m
d_0	distance travelled with zero pull, m
db	dry basis
dim	dimensionless
D	average annual depreciation, £
D_r	duration of rainfall, h
D_v	digestible organic dry matter, dim.
DM	silage dry matter, %
$DOMD$	digestible organic matter, dry basis
e	base of natural logarithm
E	gross energy value of the DOMD, MJ/kg
E_a	actual evapotranspiration rate, mm/day
E_p	potential evapotranspiration rate, mm/day
f	failure rate per unit time, 1/h
f	frequency, Hz
fc	specific fuel consumption, ℓ/kW h
F_h	drawbar pull, kN
$(F_h)_{max}$	maximum drawbar pull, kN
F_r	resultant draught force, kN
FC	fuel cost, £/h
FE_c	field efficiency for conventional ploughing, dim.
FE_r	field efficiency for reversible ploughing, dim.
g	acceleration due to gravity, 9.8 m/s^2
G	soil reaction force on the tyre, kN
h	depth of soil profile, mm
h	tyre section height, m
h_t	vertical distance between resultant draught force and rear wheel contact patch, m

H	shearing force, kN
H_i	information, bit
i_{apr}	annual percentage rate of interest, %
i_i	investment interest rate, %
i_ℓ	loan interest rate, %
i_n	net interest rate, %
i_r	real interest rate, %
i_1 to i_{12}	monthly heat indices
I	difference between amount lent and total of instalments (total interest charges), £
I	interest charge, £
I_h	annual heat index
I_{L1}	livestock intake, kg
I_{L2}	livestock intake, g/kg liveweight
I_m	antecedent soil moisture index, mm
I_t	price index for tractors, dim.
j	inflation rate, %
k	rate constant, dim.
k_g	geometric spacing factor, dim.
kd	drying coefficient
K	repair constant, dim.
K_c	cohesive coefficient, dim.
K_d	soil dryness correction factor, dim.
K_{fc}, K_{sat}	hydraulic conductivity at field capacity and at saturation, mm/day
K_g	crop stage of growth correction factor, dim.
K_ℓ	day length correction factor, dim.
K_r	simple reaction time, s
K_s	soil surface cover correction factor, dim.
K_{t1}, K_{t2}	early and late timeliness coefficients
K_w	wet day correction factor, dim.
K_ϕ	frictional coefficient, m
l_f	length of furrow, m
l_1	tractor wheelbase, m
l_2	horizontal distance between tractor/implement centre of gravity and tractor rear axle, m
L	amount of loan, £

m	repair exponent, dim.
m_0, m_1	initial and final swath or grain moisture content, % db
m_a	actual soil moisture content, mm
m_{db}, m_{wb}	moisture content, dry basis or wet basis, %
m_e	equilibrium moisture content, % db
m_{fc}, m_{sat}	soil moisture content at field capacity and at saturation, mm/300 mm profile
m_g	moisture content of standing grain, % db
m_p	soil moisture content on previous day, mm
mi	mildew, %
M	annual mortgage payment, £
ME	metabolisable energy, MJ/kg
MI	cumulative mildew index, % mildew × day
MN	tyre mobility number, dim.
n	machine age, yr
n	age exponent, yr
n	duration of dry period, day
n	number of components, dim.
n	number of part yearly intervals, dim.
n_i	number of instalments per year, 1/yr
N	number of ratios, dim.
N	period of ownership, yr
N_a	number of alternatives
N_A	ammonia-nitrogen, g/kg N
N_b, N_c, N_t	number of bodies, coulters, tines
N_i	total number of instalments, dim.
N_ℓ	number of daylight hours, h/month
N_r	number of opening ridges, dim.
N_s	number of sunshine hours, h/month
N_0	number of freezing and thawing cycles
N_2	number of two-day dry periods
NPV	net present value, £
NPV_m	total present mortgage cost, £
NPV_r	total present repairs and insurance cost, £
NPV_s	total present resale cost, £
oc	oil consumption, ℓ/h
OC	oil cost, £/h

p	probability of recurring weather sequence
p_c	crop price, £/t
p_f	fuel price, £/ℓ
p_o	oil price, £/ℓ
P	equivalent p.t.o power required, kW
P_i	axle power input, kW
P_{max}	maximum p.t.o. power, kW
P_o	tractive power output, kW
P_{ref}, P_{rms}	reference and root mean square sound pressure, N/m^2
$PP, PP_c, PP_d,$ PP_p, PP_{2wd}	initial purchase price, of cultivators, of drills, of ploughs, of two-wheel drive tractors, £
PWP	permanent wilting point, % w/w
q	purchase price coefficient
Q	wheel torque, kN m
Q_d	drainage from a soil segment, mm
Q_e	evapotranspiration from a soil segment, mm
Q_p	precipitation or irrigation water on a soil segment, mm
Q_{po}	precipitation retained by the soil with no runoff, mm
Q_r	runoff from a soil segment, mm
Q_{RR}	torque overcoming rolling resistance of driving wheels, kN m
r	purchase price coefficient
r_c	component reliability, %
r_m	mean turning radius, m
r_r	rolling radius of tyre, m
R	annual repair cost, £
R_c	clay ratio, dim.
R_f	soil reaction under front wheels, kN
R_{gs}	grain: straw ratio, dim.
R_p	pattern time ratio, dim.
R_r	soil reaction under rear wheels, kN
R_s	system reliability, %
R_u	power utilisation ratio, dim.
RR	rolling resistance, kN

s	purchase price coefficient
s	wheelslip, dim.
s_{opt}	optimum wheelslip, %
s_{sp}	travel reduction, dim.
S	resale value, £
S_c	surface cover, %
Si	silt content, %
SPL	sound pressure level, dB
t	time, s
t	time, h
t_1, t_0, t_2	start, optimum and finish time, calendar day
t	tax rate, %
t_i	tractor idle travelling time, h
t_m	headland manoeuvring time, h
t_p	tractor ploughing time, h
t_r	mean choice reaction time, s
t_{t0}	turning time, s
t_t	tractor turning time, h
t_w	walking time, h
$\triangle t$	time interval, h
T_a	air temperature, deg C
T_{a1} to T_{a12}	mean monthly air temperature, deg C
T_p	period, s
T_s	soil temperature, deg C
TC	timeliness cost, £
U	accumulated use, h
v	velocity, m/s
v	travel speed, m/s
v_{sp}	forward velocity at zero pull, m/s
v_0	theoretical forward velocity at zero slip, m/s
V	operating speed, km/h
V_i	speed of idle travel, km/h
V_{max}	maximum travel speed, km/h
V_{min}	minimum travel speed, km/h
V_p	speed of ploughing, km/h
V_w	speed of walking, km/h
V_0, V_1	typical and actual operating speeds, km/h

w	inflated discount factor, dim.
wb	wet basis
w/w	weight for weight
W	applied load, kN
W_a	total width of lands, m
W_c, W_f	coulter spacing; furrow width, m
W_{dm}, W_w	weight of dry matter or water, kg
W_e	effective operating width, m
W_f	dynamic weight on front wheels, kN
W_h	width of headland, m
W_i	static weight of implement carried on tractor, kN
W_L	animal liveweight, kg
W_m	machine processing width, m
W_p	width of plough, m
W_r	dynamic weight on rear wheels, kN
W_{rs}	static weight on rear wheels, kN
W_t	total tractor weight, kN
x	distance between markers, m
x_h	time of harvesting, day
x_r	time of ripening, day
X	displacement, m
X_h	harvest duration, day
X_r	spread of ripening, day
X_1	delay period, day
X_2	duration of harvest from time of crop ripeness for combine harvesting, day
y	yield loss, %
y_{ov}	overall yield loss, %
y_d	daily yield loss, t ha^{-1} day^{-1}
y_g	gale loss, %
y_s	shedding loss, %
y_t	threshing loss, %
Y, Y_o, Y_{av}	crop yield, at optimum time, and average over a timespan, t/ha
Y_I	yield increase, t/ha
Y_L	yield loss, kg/ha
Y_T	total or cumulative yield loss, t

Z	specific ploughing resistance, kN/m^2
α	drainage rate exponent
β	evaporation rate exponent
γ	soil specific weight, kN/m^3
γ_d	soil bulk density, kg/m^3
δ	tyre deflection under load, m
η_f	field efficiency, %
η_t	tractive efficiency, dim.
$(\eta_t)_{max}$	maximum tractive efficiency, %
θ	soil moisture content, % w/w
θ	angle of resultant draught force to horizontal, deg.
θ_t	soil moisture tension, atmos
λ	mouldboard tail angle, deg.
ν	cumulative vapour pressure deficit, $kN\ h/m^2$
π	angle of logarithmic spiral, rad.
$\emptyset$	angle of internal shearing resistance, deg.
$\emptyset$	soil water pressure head, mm
ω	rotational speed, rad/s

1

FARMING PATTERNS

OUTLINE

Agricultural output; Food surpluses and shortages; Commodity prices and production subsidies; Dietary changes; Food quality; Alternative land uses; Access to the countryside; Landscape planning; Environmental pollution; Production efficiency; Energy budgets; Animal welfare; Agricultural work force; Holding size; Tractor ownership; Average tractor power; Machinery management.

APOSTROPHE – THE ROADMENDER

'The swift stride of civilisation is leaving behind individual effort, and turning to the Daemon of a machine. To and fro in front of the long loom, lifting a lever at either end, paces he who once with painstaking intelligence drove the shuttle. *Then* he tasted the joy of completed work, that which his eye had looked upon, and his hands had handled; now his work is as little finished as the web of Penelope. Once the reaper grasped the golden corn stems, and with dexterous sweep of sickle set free the treasure of the earth. Once the creatures of the field were known to him, and his eye caught the flare of scarlet and blue as the frail poppies and sturdy corncockles laid down their beauty at his feet; now he sits serene on Juggernaut's car, its guiding Daemon, and the field is silent to him.

Now the many live in the brain-sweat of the few; and it must be so, for as little as great King Cnut could stay the sea, so little can we raise a barrier to the wave of progress, and say: "Thus far and no further shalt thou come."

What then? This at least; if we live in an age of mechanism let us see to it that we are a race of intelligent mechanics; and

> if man is to be the Daemon of a machine let him know the
> setting of the knives, the rise of the piston, the part that each
> wheel and rod plays in the economy of the whole, the part that
> he himself plays, co-operating with it.'
>
> (Source: Fairless)

1.1 INTRODUCTION

Highly mechanised agriculture in Europe and North America
faces tighter margins and increasing restrictions as society wres-
tles with the immorality of food surpluses in a world, beset by
malnutrition and famine, where those in the greatest need have
neither the foreign exchange to purchase food nor the appropriate technology to modify their subsistence farming.

The attainment of self-sufficiency in food production has been
associated with substantial changes in consumer trends and in
public opinion. Dietary adjustments and callisthenics are being
encouraged to compensate for the general reduction in heavy
manual work and the almost total reliance on vehicular transport.
Conservation of the environment is beginning to take precedence
over the strategic importance which has been traditionally
accorded to farming to survive a siege economy in times of
national need. The political influence of the farming community
is being eroded as their numbers decline through the amalgamation of holdings, and through the use of larger machines.

In addition to their still essential role as food producers,
farmers are seen increasingly as guardians of the landscape. The
urban population, with greater leisure and mobility, are seeking
not only more access to the countryside but also more resources
for the maintenance of that amenity.

Despite these changes in attitudes, people must eat. There is
a continuing need for advanced machinery to meet higher
production targets as profits fall and as more land is used to
satisfy the demand for timber. The more stringent quality standards for the 'fast food' and the 'frozen fresh' industries place
greater emphasis on operational timeliness and maximum
machine reliability.

Exploiting the full potential of expensive, high-output equipment involves better machinery management as well as better
business management. It is no longer sufficient to match indi-

vidual machines by trial and error. The financial penalty of an incorrect purchase is considerable and the duration of the mistake more protracted as machines are replaced less frequently.

Whilst the complete analysis of mechanisation systems is sufficiently complex and time consuming to require the application of computer programs, an understanding of the practical basis of the analysis is essential for the credibility of the output results. In addition, the practical data on separate aspects of labour, machinery and economic performance represent a valuable source of reference for improving weak links in existing farm machinery operations.

1.2 AGRICULTURAL OUTLOOK

Following the food shortages of the Second World War, the agricultural industry in Europe has been encouraged to increase output by offers of production subsidies and guaranteed commodity prices. Although the land area devoted to agriculture has not increased, crop yields and animal output have improved to a level where the production of many commodities is in excess of self sufficiency (Table 1.1). The resultant surpluses must either be stored indefinitely or sold on world markets at discounted prices. Both solutions increase food prices for the consumer and there is mounting political pressure to achieve a finer balance between food production and consumption.

It is important to recognise that yield and efficiency are not synonymous: yield is a measure of output, whereas efficiency is the ratio of yield output to resource input. It is undeniable that the yield of crops and livestock products have increased dramatically in British agriculture. Average yields of wheat have more than doubled in thirty years, whilst those for barley, oats, potatoes, eggs per bird and milk yield per cow have increased by around 80 per cent (Table 1.2). These substantial yield increases do not automatically infer greater efficiency because they have been associated with many other changes in the balance of resources used in the production systems and in the scale of operations. The reduction in number of people employed in agriculture has been accompanied by an increase in the number employed in the downstream food processing and distribution

Table 1.1　The degree of self-sufficiency in various agricultural commodities in the EEC for 1983–84

Commodity	Self-sufficiency, %	
	EUR 10	UK
Soft wheat	117	98
Barley	108	138
Oats	94	97
Potatoes	99	87
Sugar	123	55
Vegetables	98	61
Fresh fruit	83	25
Wine	101	0
Oleaginous fruits and seeds	45	38
Vegetable fats and oils	38	23
Skimmed milk powder	133	229
Cheese	107	74
Butter	147	77
Eggs	103	98
Beef and veal	105	86
Pork	102	71
Poultry meat	111	96
Sheep and goat meat	75	72

(*Source: Yearbook of agricultural statistics 1985*, EEC 1986).

industries. The relatively low direct fuel consumption for field operations must be balanced against the high energy requirements for manufacturing nitrogen fertilisers and the petroleum base for agrochemicals. Even in economic terms, the investment in land improvements, buildings and machinery have increased farming indebtedness to a level which cannot be sustained by falling commodity prices.

Table 1.2　Improvements in output from British agriculture over a 30-year period

Year	Yield					
	Wheat, t/ha	Barley, t/ha	Oats, t/ha	Potatoes, t/ha	Eggs per bird	Milk per cow, ℓ
1950–52	2.7	2.7	2.3	19.6	142	2703
1960–62	3.9	3.4	2.8	22.4	191	3535
1970–72	4.3	3.7	3.7	29.2	226	3943
1980–82	6.0	4.6	4.3	35.7	250	4789

In response to lower product prices, the farmer really has three options:

o **take land out of use;**
o **find alternative products;**
o **alter the level of inputs and outputs.**

Individually, most farmers are unenthusiastic about taking *land out of production* because of the effect not just on their own livelihood which can be protected by compensation, but also on the whole rural infrastructure when employment is decreased. The land most likely to suffer first is already marginal with natural deficiencies through higher altitude locations, north-facing aspects, or poor drainage, or with a combination of all three. During periods of over-production and falling commodity prices, changes in use occur when farmers cease trading, the land prices for marginal farms falling faster than for high quality land. Taking land out of agriculture does not necessarily imply that it will be abandoned, and many would argue that some redistribution of land to forestry, conservation and recreation is both desirable and overdue. This applies to lowland arable farms through the establishment of small scale hardwood plantations and the adoption of coppicing techniques as well as to the uplands through an expansion of agroforestry with softwoods.

In any climatic region, there is only limited scope for turning to *alternative agricultural products*. Over much of Britain, the most successful and noticeable alternative crop is oilseed rape with its dominant splash of yellow flowers transforming the countryside for a time into a patchwork which irritates and delights almost equal proportions of the population! Other crops have been tried with less widespread adoption. Latitude is an obvious restriction for the sunflower crop, whereas the high processing costs contributed more to the failure in re-establishing an initiative with the flax crop in areas of Scotland which had witnessed substantial production half a century earlier.

Exclusively arable systems are capital intensive, and generally maximise cash income for a given level of quality of land and management, through the high input of nitrogen fertiliser and crop protection chemicals to sustain crop yields at close to the biological maxima. Rotational husbandry in conjunction with a livestock enterprise – the traditional mixed farming system – offers a *lower input alternative* but at reduced profitability.

Differential subsidy support could improve the relative economic importance of low input farming by acknowledging the community benefits of lower nitrates in drinking water, the reduction in straw disposal by burning, and so on. A balanced cropping pattern provides opportunities for integrated pest/disease/weed control procedures, and the effluent from the animal enterprise allows recycling of plant nutrients and soil conditioners, thereby allaying public concern about the dangers of inorganic chemical residues impairing food quality. In order to avoid surpluses of animal products, the reintroduction of livestock on arable farms must be associated with less intensive production on livestock farms, trends which would both placate the animal welfare lobby and meet changing dietary demands.

1.3 CONSUMER TRENDS

The consumer is becoming much more aware of the health aspects of food composition and purity. There is particular concern about the adverse effects of relatively high levels of sugar, animal fat and salt intake, as well as the relatively low levels of fibre and fruit intake. The necessity for the growing number of additives in processed foods is also under scrutiny. Some of these additives are included as preservatives, some as colouring or flavouring, some to maintain or increase the water content and yet others as minerals and vitamins. Whilst there is a strong demand for convenience foods which undergo some compositional changes during processing, there are those who are suspicious of processing in any form and who are prepared to search for texture, flavour and purity by paying a premium for products direct from the farm or from health food shops.

1.3.1 Nutritional awareness

In most industrialised communities, deficiency diseases have been virtually eliminated through the adoption of the concept of the 'balanced diet'. By ensuring that a variety of foods are consumed, it is more likely that one item rich in a particular nutrient would 'balance' the lack of this nutrient in another food. Even so, the introduction of a greater variety of foods is unlikely to combat the increase in diseases of affluence, such as coronary heart disease, bowel disorders, and obesity. The present need is to alter the proportions of the food items consumed and to introduce new food products whilst still maintaining an adequate variety of foods – encapsulated in the term, 'prudent diet'.

The body makes first call on the diet for energy. This can be supplied by any proportion of the three macronutrients, namely: carbohydrates, proteins and fats. As most diets of the world are largely based in cereal or root crops, the population relies on starches for 50 to 75 per cent of the energy requirement. These staple foods also contain between 10 per cent protein (cereals) and 20 per cent protein (protein-rich foods), so that the diet invariably contains 10 to 15 per cent protein – about double the minimum physiological requirements when hunger is satisfied. The remaining 10 to 40 per cent of the energy comes from fats.

Within these broadly acceptable proportions of macronutrients, there are important dietary goals which require a change in foods rather than nutrients – less of the fats and simple carbohydrates (the sugars) and more of the complex carbohydrates (the high fibre foods).

Fat The consumption of saturated fatty acids and of fat should be reduced by one quarter to nearer 30 per cent of total energy intake, thereby increasing the ratio of polyunsaturated fatty acids to saturated fatty acids to appoximately 0.45. Less fat lowers the level of cholesterol in the blood and reduces the incidence of coronary heart disease. This dietary change involves breeding leaner meat, reducing the fat content of processed meat products and changing to low butterfat milk.

Sugar The current consumption should be cut by one half to 10 per cent of the total energy intake. Energy without any accompanying nutrients – empty calories – encourages obesity, defined as 20 per cent overweight, and dental decay.

Fibre The present intake should be increased by about 50 per cent to 30 grams a day. The shift from foods high in fat and sugar to whole foods involves increasing the consumption of bread – particularly wholemeal – vegetables and fruit.

Salt Only about 1 gram a day is essential and the average daily consumption of nearly ten times this amount should be halved. Excess salt is linked with hypertension. Better labelling of processed foods might discourage the inclusion of both salt and sugar as taste intensifiers.

1.3.2 Food quality

Food *quality* is an elusive factor whose importance grows in proportion to the affluence of the consumer.

- **Quality** is the resultant of several relative values which, considered together, determine the acceptability of the product to the consumer.

The seven basic qualities of any fruit or vegetable product are:

- o **freshness;**
- o **colour;**
- o **flavour;**
- o **texture;**
- o **size and uniformity;**
- o **nutritional value;**
- o **freedom from defects.**

These different qualities of a particular product do not equally contribute to the overall quality image, size being least important.

The image of *freshness* is closely associated with the quality of *nutritional value* but the term is not easy to define. Many of our fresh foods are far from fresh by the time they reach the retail outlets unless precise steps are taken to inhibit deterioration, whereas the British Frozen Food Federation has as its slogan: 'You can be sure it's fresh if it's frozen'. True freshness is related to proximity of harvesting; the sooner the natural deteriorative processes are arrested by freezing or canning, the fresher the commodity. The cool or chill chain is an extension of this principle for slowing the ageing processes of fresh produce and improving its shelf life. The psychological impact of freshness is fully recognised and large retail outlets often display such produce to maximum advantage at the entrance to the store.

Colour has a major effect on consumer preferences and needs to match the anticipated flavour: 'the consumer eats through her eyes'. The sensory perception of colour can be sub-divided into hue, uniformity and brightness, each varying with crop maturity as well as with cultivar. Naturally pale hues are preferred when artificial colour is used to enhance product appearance but increasing opposition to additives is encouraging a change to produce with naturally darker hues. Although complete uniformity of colour is difficult to achieve in a biological material, brightness should enable the product to sparkle.

Flavour is a subtle combination of the four basic tastes which can be detected by the human palate – bitter, sweet, sour and salt – and over 10 000 odours. Desirable flavours induce the sensation of appetite, whilst taints cause food to be rejected. These taints, or foreign flavours, may be caused by pesticide residues, over-maturity, delay between harvesting and processing, faulty preparation, or subsequent contamination. Within the acceptability range, distinctive flavours provide an identifiable brand image for processed foods through the addition of artificial flavouring, sugar or salt. Consistency of product flavour is maintained by using a team of tasters to monitor the subtle nuances of flavour and record the differences in quality note on a sensory analysis profile star (Fig. 1.1).

Texture provides an assessment of the tenderness, ripeness and wholeness of produce. Vegetables become tougher as they mature, fruit become softer as they ripen, and celery must remain crisp. Textural characteristics are best appreciated in the mouth. Some compensation for the inevitable loss of natural texture through cooking or processing is possible by adjusting the time of the harvesting.

Size and uniformity are graded by length, circumference, diameter, width, volume and weight. Whilst size is not a quality characteristic in itself, smallness often has flavour and textural connotations, much as new potatoes. Close grading is also being used increasingly as a brand image linking a particular uniform size of fresh produce with a specific retail outlet.

Nutritional value is not confined solely to the chemical composition of the agricultural commodity itself, but also includes the more emotive health issues associated with the use of agrochemicals for crop production and protection, as well as the inclusion of food additives during processing. Although excessive use of preservatives merely to extend shelf life is to be discouraged,

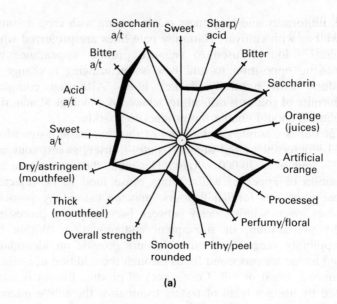

(a)

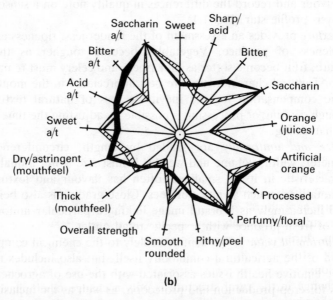

(b)

Fig. 1.1 Examples of sensory analysis profile stars for: (a) a single sample of orange juice; (b) two juice samples superimposed so that the differences in quality note can be clearly seen (from Arthey, 1986).

total prohibition would cause a rapid increase in food poisoning. Equally, the adherents of organic farming must adjust their perception of other quality standards to resolve the conflict between 'residue free' and 'defect free'.

Freedom from defects is of immense importance to the producer and the consumer alike. The ravages of both pests and diseases are never totally eliminated, despite the most dedicated spray control programmes. Environmental irregularities create water and nutrient shortages which result in crop growth defects. Some genetical changes in plants impair marketability. Extraneous matter from the growing medium (for example, stones and grit), from the plant or adjacent weeds, and from agrochemical residues are particularly serious. Finally, mechanical damage not only reduces the quantity of saleable produce but also impairs the condition of the final product through internal bruising which may not appear until after processing.

Improving the quality of agricultural products is as much dependent on effective harvesting, processing and distribution systems as it is on the intrinsic characteristics of the crop itself.

1.4 CONSERVATION AND THE COUNTRYSIDE

Encouragement of recreation and greater availability of personal transport have generated more strident demands to improve access to the countryside and to protect the landscape from hedge removal, tree felling and closure of footpaths. Conservationists seek to protect wildlife habitats by preserving wetlands, woodlands and heath, and to protect rare plant species by legally preventing land improvement at Sites of Special Scientific Interest. The spread of commuters into agricultural communities has also provided a stronger lobby to curb the worst excesses of modern agricultural practices which cause nitrate pollution of water supplies, and atmospheric pollution from the obnoxious odours during slurry spreading and from the smoke and ash during straw burning. Although the imposition of some additional restrictions on the farming industry appears inevitable, the change in the rural structure is only a change of emphasis; the urban interests, the wildlife interests and the farming interests are not in total conflict with each other.

1.4.1   Landscape planning and land use

Landscape is primarily determined by geology, topography and climate. Agriculture modifies the scenic expression of these fundamental factors but requires only minor sacrifices in operational efficiency to maximise the aesthetic value of the countryside. The geological diversity of Britain ensures that many farms contain small areas of poor soil – low-lying patches of waterlogged clay, or caps of gravel, or thin soil, or steep banks – where cultivation is difficult. Within this geological framework, the development of a complex social infrastructure has also isolated numerous odd-shaped pockets of land at the junctions of fields with roads, railways and water courses. It is a small loss to farming and a great gain to the landscape and to conservation if these areas are planted with trees or left as rough grazing.

The visual impact of the landscape is difficult to assess because it differs not only between individuals, but also between one mood and another in the same individual. Wide views are valued for their sense of freedom, the enclosed landscape of hedged fields and woodlands for its sense of intimacy and safety, ruggedness for its drama, and calm estuaries for their tranquillity. Rather than just giving an appreciation of pictorial composition, the instinctive recognition of the biological balance of an organic structure seems to give a sense of repose. The atavistic desire for security, whereby people prefer to sit at the edge of a wood rather than either inside or in the open, is satisfied by the functional network of hedges which are primarily intended for the enclosure of land into management units, for shelter and for the confinement of stock. Significantly, however field boundaries provide the diversity of habitat which attracts other species, as well as man, and provide corridors for wildlife to travel across areas of intensive crop production from one eco-system to another. Despite the reduction in hedgerows to accommodate larger machines, the situation only becomes critical where hedgerow densities fall to below 200 or 300 metres a hectare, where few small woods exist in the locality, and where the network of wildlife corridors is not sustained – albeit on a larger scale of mesh – along motorways, railways and waterways. These arteries traverse field boundaries, shelter belts, rivers and streams to link the lowlands with open country.

Landscape patterns also contain a blend of visual contrasts between horizontal fields and vertical trees, between regimented rowcrops and natural climax vegetation and between water meadows and irrigation lagoons. Traditional groups of farm buildings, built of local materials, form strong focal points in the pattern of the landscape, often seeming to draw the network of hedgerow trees together into a sheltering group. New buildings, however, need not detract from the image of the country scene, provided that they are well proportioned, well sited and of the right colour and texture. Even a tall silo or an aero generator, whose very size and strangeness cause initial alarm, can contribute a key vertical element to the composition, bearing comparison in time with a church tower rising above a hamlet or an old windmill. The horizontal scale relation between the buildings and the surrounding expanse of land is also vitally important, for a landscape crowded with structures (seen at its worst in fields covered with battery hen houses) ceases to be pastoral and is visually transformed into an industrial scene.

1.4.2 Environmental pollution

Environmental pollution includes terrestrial, aquatic and atmospheric effects, often in combination. Soil erosion is exacerbated by the removal of hedgerows which are capable of influencing the wind over a total distance of 40 times its height, three-quarters on the leeward side. The likelihood of soil loss is all the greater with the popularity of winter cereals because the slower plant growth during the establishment period leaves the soil devoid of plant cover over the autumn equinox when the frequency of gales and the probability of heavy rain are high in a maritime climate. Public indignation at the resultant dust clouds and mud-covered roads masks the serious implications to farming through the depression of crop yield, through extra cultivations to eradicate rills, and through the leaching of fertiliser directly into water courses.

The favourable seasonal distribution of rainfall, augmented by scheduled irrigation, normally permits field crops to respond fully to fertiliser and agrochemical inputs, and ensures the success of intensive farming patterns. Although the manufacture of fertiliser accounts for a significant fraction of the total energy used in agriculture, the financial cost of fertiliser is still relatively low. As only perhaps 10 per cent of the nitrogen applied to farm crops is absorbed by the plants themselves, the escalation in the use of fertiliser has caused high levels of nitrates in water courses and in acquifers. These abnormal concentrations can result in algal blooms which make the water unsuitable for industrial processes. The nitrate content of drinking water should not exceed 50 mg/litre to minimise the risk of inducing methaemoglobin-aemia in very young bottle-fed babies. Fortunately, the under-standing of the seasonal dynamics of nitrogen transformations in the soil, as influenced by soil type, temperature and rainfall, is improving rapidly. This more comprehensive assessment of the amount of mineral nitrogen likely to be available to, and needed by, crops at different growth stages leads to enhanced accuracy of application rates and of application timing, with lower wastage.

The handling, storage and spreading of animal wastes cause offence not only to the farm worker but also to the surrounding public because of the smell, the disease hazards, and because of the risk of pollution to water courses. Stocking densities of inten-sive livestock enterprises may well be restricted to allow effluent disposal on locally available land at acceptable application rates for crop production and on unsaturated soils to avoid surface runoff and soil damage. Shorter periods suitable for effluent disposal increase the demand for storage lagoons which require careful siting to limit both danger and nuisance. In order to avoid the formation of fine aerosols which transport obnoxious odours on the wind to the distaste of everyone to leeward, the use of scrubbers mitigates ammonia emissions from ventilation systems of factory farms, and slurry injection into the soil minimises the manure smells during disposal operations.

The change from straw bedding systems of livestock housing on upland farms has increased the straw surplus on arable farms. The simplest method of disposal through burning is not the most environmentally effective for a number of reasons. A sudden shift in wind direction can cause havoc to traffic enveloped in dense smoke; a sudden increase in wind strength can create a major fire risk to adjacent hedgerows, woodlands, neighbours' crops, and even property; but, worst of all, are the smuts caused by the wind-blown ash during periods of prolonged drought. Incorporation of chopped straw into the soil during the cultivations for the following crops provides a suitable alternative to straw burning but requires more complex machinery to mix the straw and soil, additional nitrogen to decompose the straw once mixed, and better crop hygiene to prevent diseases from using the residues as a host to bridge between growing seasons.

1.5 PRODUCTION EFFICIENCY

Plants are a rather inefficient method of harnessing solar energy, only fixing some 1 to 2 per cent of the incident solar energy in their tissues. In total, however, the quantities of energy available are enormous and much is wasted. Of more immediate concern is the ratio of the food energy output to the production energy input. Subsistence farmers can produce an adequate output of food energy for a very modest input of human energy and can attain energy ratios of 15 to 20 for most crops. These high figures epitomise the life styles of such people: they do not need to work very hard to provide their basic needs but they are profligate in the use of another resource, namely, land.

Intensive agriculture provides a complete contrast. The improvements in crop production efficiency have been achieved at a cost of very large inputs of energy in other forms – the fossil fuel subsidy. These energy reserves from past solar radiation are being consumed for the production of fertilisers, agrochemicals and machinery, for fuel, and for the transportation and processing of food products beyond the farm gate. Although the energy ratio falls towards unity (and even sometimes below it), there is nothing inherently wrong in this, provided that the contribution of non-renewable energy sources is properly acknowledged and the rate of depletion of the reserves is adequately controlled.

Energy budgets are equally important in livestock systems but, in addition, production efficiency is sometimes achieved at the expense of animal welfare. Greater understanding of, and attention to, animal behavioural patterns should lead to better housing layouts, fail-safe feeding and ventilation systems and gentler livestock handling procedures.

1.5.1 Energy budgets

The direct use of energy as fuel on British farms is under 2 per
cent of the national fuel requirement. If the upstream use of
energy for inputs of fertilisers and so on is included, the total
energy input of 4 per cent for agriculture is still modest compared
with other heavy industries. The main contributors to the agricultural demand for support energy are the direct use of fuel (24
per cent), the manufacture of fertilisers (23 per cent), off-farm
processing of animal feed (15 per cent), the manufacture of
machinery (14 per cent), and the direct use of electricity (9 per
cent). The support energy for miscellaneous items, such as buildings, agrochemicals and distribution services, make up the
remaining 15 per cent.

Beyond the farm gate, additional support energy is required
for processing the agricultural commodities and distributing
edible products to the retail outlets. In the energy budget for
bread production, for example, milling, baking and distribution
account for about one third of the total energy cost, whereas
transportation of sugar beet to the factory gate and the sugar
refining process account for over two thirds of the total support
energy.

The energy budgets vary considerably for different commodities (Table 1.3). The support energy for potatoes is almost three
times that for wheat, whilst the energy output per hectare is
similar. The manufacture of nitrogenous fertilisers is particularly
energy intensive. Nitrogen represents almost a third of the
support energy for these arable crops and becomes a dominant
factor in intensive grassland systems.

Estimates of the agricultural use of support energy are given
in Table 1.4. The support energy inputs and metabolisable energy
outputs for the various commodities are related to the annual
land area required to sustain production. The energy ratios are
generally higher than unity for crops and well below that figure
for animal products. Less energy is required to produce protein
from arable crops than from animals but, despite their poorer
conversion efficiency, animals do produce edible protein from
material which could not be eaten by man directly.

The lowest energy ratio is obtained for tomatoes because of
the very high heating requirements. Such specialist crops cannot

Table 1.3 Energy budgets for two agricultural commodities

Energy contribution	Energy, MJ/ha	
	Wheat	Potatoes
Inputs		
Fertiliser		
Nitrogen at 77 MJ/kg	7 430	13 100
Phosphate at 14 MJ/kg	665	2 530
Potash at 8.3 MJ/kg	322	1 610
Herbicides at 106 MJ/kg	212	1 380
Seed	875	4 730
Fuel for field operations	2 330	5 440
Grain drying	5 760	—
Grading and storage	—	20 200
Machinery depreciation and repairs	2 030	2 920
	19 600	51 900
Outputs		
Yield of 4.22 t/ha at 15% moisture content	61 000	—
Yield of 27.50 t/ha	—	69 300
Energy ratio	3.11	1.33

(*Source*: White, 1975)

be justified solely on the energy flow. Assuming that market demand for this type of commodity is maintained because of other less quantifiable benefits derived from eating a varied diet, preference for using energy to grow a protected crop in an adverse environment instead of for transportation from a more favourable climatic zone depends on the monetary economics associated with the various choices.

There are also opportunities for increasing the energy ratios in crop production. Some scope still remains to reduce the support energy through better design of farm machines, more effective application techniques for plant protection to reduce agrochemical inputs and greater utilisation of agricultural by-products and wastes with high residual energy contents. In the longer term, the output energy level may be enhanced through the development of crops and cultural practices which would lead to the higher net conversions of solar energy into plant dry matter. This could include the benefits of increased radiation interception

Table 1.4 Estimates of agricultural use of support energy

Product	Support energy input, GJ/ha yr	Metabolisable energy output, GJ/ha yr	Energy ratio	Protein output, kg/ha yr	Support energy for protein, MJ/kg
Wheat	19.6	61.0	3.11	435	45
Barley	18.1	60.6	3.36	310	58
White bread	31.7	47.1	1.48	368	86
Potatoes	52.0	69.3	1.33	460	113
Sugar beet	109	82.5	3.28	—	—
Sugar from beet	109	82.5	0.76	—	—
Carrots	25.1	32.5	1.30	234	107
Brussels sprouts	32.4	10.9	0.34	296	109
Tomatoes (under glass)	1300	62.0	0.05	945	1360
Milk	17.0	12.0	0.70	145	118
Beef	10.6	2.4	0.23	31	348
Pork and bacon	18.0	11.4	0.63	76	238
Lamb and mutton	10.1	2.5	0.25	22	465
Poultry (eggs)	29.4	4.3	0.15	145	203
Poultry (broilers)	23.6	7.1	0.30	129	184

(*Source:* White, 1975)

from earlier seasonal establishement of crops or better growth at lower temperatures, and the likelihood of manipulating the genetic composition of crop plants to produce cultivars which exhibit higher intrinsic photosynthetic efficiencies.

1.5.2 Animal welfare

There is public disquiet about all forms of 'factory farming', and some pressure groups are pledged to abolish totally the housing of poultry in battery cages for egg production, tethered sows, and close confinement of veal calves. Attempts to ban livestock exports are a further indication of the growing concern for animal welfare and this aspect has implications for the movement and handling of livestock generally. It is inappropriate to recommend special procedures governing the transportation and handling of livestock across national frontiers without including equally arduous journeys in the country of origin.

Codes of Practice for the management and handling of livestock provide a balance between the production demands and the animal welfare issues. Revised cage allowances for layers give far more space per bird. Groups of dry sows can have free access to yards through the availability of individual feed dispensers. A transponder, fitted by a neck collar to each animal, gives separate identification at the feed station and a computer-controlled system dispenses the required amount of feed.

Inevitably, these improvements increase the cost of production. There is some market acceptance of higher prices for animal welfare linked with perceived product quality. Free range eggs sell at a premium but other examples of this type are hard to find. The consumer support for animal welfare is also in evidence through the reduction in consumption of animal products. The move to a more vegetarian diet (though not exclusively so) is consistent with healthier living, is conservative in the use of land and energy resources, and does not offend those concerned with animal welfare.

1.5.3 Labour demand and size of holding

In the United Kingdom over the past thirty-five years, the number of full-time agricultural workers (including family) has

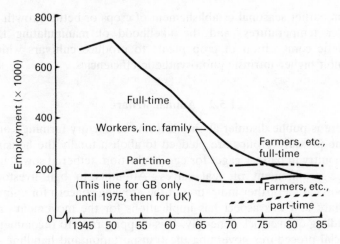

Fig. 1.2 The steady decline in the numbers of full-time agricultural workers in the UK since 1950.

fallen steadily from a peak of 700 000 to under 200 000, and is still declining (Fig. 1.2). This decline reflects the elimination of drudgery and the reduction in heavy manual activities through the greater use of mechanisation. The numbers of part-time and casual workers have remained virtually constant because there are still harvesting operations which rely on squads of pickers to attain high quality standards for the fresh fruit and vegetable markets. There is no indication of the duration of the employment period for this category, but it is likely that total number of man days is declining even though the number of workers is unchanged.

Records of the number of farmers, managers, directors and partners are rather more recent. It is interesting to note that the number of part-time farmers is increasing as more and more seek to augment their dwindling incomes, particularly on the smaller farms. This trend, together with the amalgamation of farms into more economically viable units, reduces the remaining number of full-time farmers.

Between 1960 and 1984, the average size of farm holdings in the United Kingdom has more than doubled from 32 hectares to 70 hectares. This average size appears too small to match the popular image of a thriving and prosperous agricultural economy because the average is heavily biased by the high proportion of small holdings. Over 50 per cent of agricultural holdings are below 30 ha, whereas only 3 per cent exceed 300 ha. In terms of land, however, the holdings below 30 ha account for only 10 per cent of the total agricultural area compared with 30 per cent for the 100–300 ha holdings and a further 35 per cent of the total area for farms over 300 ha. Even the allocation of total area to different farm sizes requires interpretation with some degree of caution because extensive hill farms are included in the analysis.

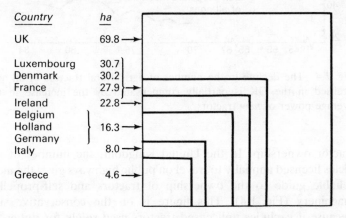

Country	ha
UK	69.8 →
Luxembourg	30.7 ⎱
Denmark	30.2
France	27.9 ⎰ →
Ireland	22.8 →
Belgium	
Holland	16.3 →
Germany	
Italy	8.0 →
Greece	4.6 →

Fig. 1.3 Average size of farm holdings over 1 ha in the EEC for 1984.

On balance, therefore, average holding size provides an acceptable guide to the structure of farming throughout the European Economic Community. On this basis, the average holding size in the United Kingdom is more than twice that for its nearest contenders (Luxembourg, Denmark and France) and over 15 times that for Greece (Fig. 1.3). Very small holdings are labour intensive and cannot acquire or justify the capital investment for mechanisation without Government assistance to encourage the formation of larger units.

1.5.4 Power and machinery

During the early stages of agricultural mechanisation, a fall in the size of the total workforce is accompanied by an increase in

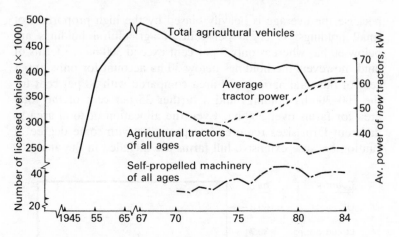

Fig. 1.4 The decline in the number of agricultural tractors of all ages licensed in the UK is partially compensated by the increase in the average power of *new* tractors.

tractor ownership. In the United Kingdom, the number of vehicles licensed annually to travel on public highways gives the most reliable guide to the ownership of tractors and self-propelled machinery (Fig. 1.4). This figure is on the conservative side because it excludes unlicensed tractors used solely for duties on the farm. Nevertheless, it is evident that the numbers of licensed tractors, of *all* ages rapidly increase at first, peak during the mid 1960s, and decline steadily thereafter. This decline in the numbers of tractors indicates that mechanisation is complete and the sales of new tractors are entering a replacement market. The market for self-propelled machinery is more recent and is still in the first phase of mechanisation – displacing smaller, trailed equipment rather than labour (Fig. 1.4).

Increasing wage costs encourage the trend towards fewer, more powerful tractors and machines. The average power of new tractors is still increasing, so that the total tractor power available has probably reached a plateau, although the statistics are too recent to be dogmatic (Fig. 1.4). There are sales variations, however, in the different tractor power categories (Fig. 1.5). Small tractors of up to 30 kW are increasing in numbers but still only represent

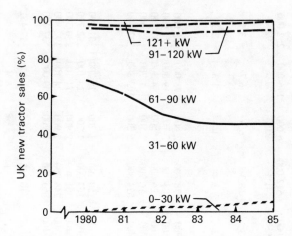

Fig. 1.5 Registration of new tractors in the UK for different power categories.

5 per cent of annual sales in the United Kingdom. At the other end of the power spectrum, there is a similar increase in sales for the 91–120 kW category. The sales of very large tractors amount to between 2 and 3 per cent. The most significant shift in buying behaviour is the transfer of over 20 per cent of the annual sales from the 31–60 kW power category to the 61–90 kW size. This market change makes the 61–90 kW power category the largest market sector, representing almost half of annual sales.

This detailed analysis of the British market provides greater insight to interpret tractor ownership data for the European Economic Community as a whole (Table 1.5). In comparison with the other member states, the United Kingdom has the lowest number of tractors per 100 ha except for Ireland where the percentage of sole ownership of tractors is still low. Another indication of larger holdings and more efficient use of tractors is given by the average number of tractors per holding. Belgium, Italy and the Netherlands, with a small average size of holding, have about 1.3 tractors per holding. West Germany with a similar average size of holding to Belgium and the Netherlands has 1.6 tractors per holding, whilst the United Kingdom has 2.3 tractors per holding.

Table 1.5 Community survey on the structure of agricultural holdings and tractor ownership

	Belgium	Denmark	France	West Germany	Ireland	Italy	Lux.	Neth.	UK
Holdings using tractors	82 000	115 000	1 170 000	826 000	149 000	1 512 000	6 000	146 000	252 000
of which in sole ownership	71%	100%	81%	98%	66%	41%	93%	71%	92%
Number of tractors in sole ownership per 100 ha	6.7	6.4	4.7	10.1	2.3	5.3	6.6	6.7	3.1
Number of tractors per holding	1.3	1.6	1.5	1.6	1.2	1.4	1.6	1.3	2.3
Holdings by number of tractors in sole ownership:									
1 tractor	51 700	59 100	623 700	447 000	84 800	478 200	2 600	78 000	88 000
2 tractors	17 600	43 000	256 800	289 600	10 500	103 100	2 100	19 800	70 900
3 tractors	2 700	9 800	52 000	59 100	2 100	26 600	500	4 300	37 700
4 tractors and over	700	2 700	18 700	11 300	1 000	16 900	100	1 700	35 600
Total	72 700	114 500	951 200	807 000	98 400	624 800	5 300	103 800	232 100
Tractors in sole ownership by power categories:									
Under 25 hp	10 700	10 700	225 500	379 800	20 100	150 100	1 200	15 700	31 300
25–34 hp	16 000	48 900	271 900	319 400	31 000	167 000	1 700	28 100	82 700
35–50 hp	38 200	70 300	480 200	374 400	47 100	301 400	2 900	71 100	230 300
Over 50 hp	33 200	57 300	408 500	181 400	19 800	244 900	2 900	24 100	192 800
Total	98 100	187 200	1 386 100	1 255 000	118 100	863 400	8 700	139 000	537 100
Agricultural area × 1000 ha	1 462	2 936	29 426	12 462	5 077	16 188	131.5	2 074	17 451

(*Source:* Eurostat 1979)

1.6 MACHINERY MANAGEMENT

Arable farming has always relied heavily on tillage implements, but it was not until the introduction of the tractor that farm power and machinery started to become a major production cost and an indispensable part of the agricultural system. The effectiveness of the mechanisation policy is determined by the management skill in matching the work output of the power and machinery complement to the time available at an acceptable level of fixed and operating costs.

During the initial stages of mechanisation, individual machines are introduced to save labour, to increase output, to improve product quality, and to reduce drudgery. These benefits are often so substantial that scant regard is given to the whole farm planning. Surplus power is a positive advantage to accommodate expanding machine capacity and diversity.

On fully mechanised farms, however, mismatched machinery systems represent a cost burden through more expensive operational charges which can only be rectified by a capital outlay to advance the date of machine replacement. As the number, size, complexity and cost of machines increase, the adequacy of *machinery management* has a major impact on farm profitability.

- **Machinery management** is the study of the selection, operation and replacement of farm machines.

Optimum machinery management is achieved when the overall profitability of the farm business is maximised. This economic goal is not necessarily equivalent to minimising machinery costs for a number of reasons. Different enterprises demand different tractor and machinery combinations and optimum utilisation may require area adjustments which are unacceptable for rotational reasons. Inadequate machine capacity may incur yield penalties from untimely operations, whereas over capacity may introduce the risk of greater soil damage due to the additional weight of the large equipment.

Although optimisation of the power and machinery requirements for diverse enterprises is a complex problem, it can be examined in three stages:

○ **maximum tractor power demand;**
○ **tractor fleet size;**
○ **tractor power mix.**

The *maximum tractor power* on an arable farm is largely dictated by the draught demand for primary tillage. *Tractor fleet size*, on the other hand, is governed by the need for simultaneous operations with individual machines, for example during crop establishment and by the transport requirements during the spring and autumn peaks. The *tractor power mix* identifies whether all the tractors in the fleet are of the same, or of varying, power sizes. These three components of tractor and machinery selection are seldom scrutinised with the same attention to detail. Power always predominates and an acceptable solution is approached by an intuitive process of replacing individual machines whenever a 'bottleneck' becomes too much of a nuisance.

Opportunities now exist to augment managerial experience with analytical procedures to obtain a more accurate estimate of tractive performance from soil, weather and tyre data, and to identify the optimum machinery complement by evaluating the costs for a whole range of operational schedules. These procedures provide a more objective appraisal of operator productivity, machine performance, available workdays and machinery operating costs. These factors, of course, combine to create a huge number of options which are more readily handled by a computer to identify a limited number of effective solutions. Nevertheless, the practical data on individual aspects of *men, machines, money and management* form a useful source of information for improving the management of existing machinery operations.

1.7 SUMMARY

Highly mechanised agriculture faces tighter margins and increased restrictions to eliminate over-production, to accommodate dietary changes and to meet the public demands for greater access to the countryside.

There is a continuing demand for more timely operations and better processing techniques to improve food quality, whose importance grows with the affluence of the consumer.

Agriculture requires only minor sacrifices in operational efficiency to maximise the aesthetic value of the countryside.

Soil erosion, nitrate pollution, obnoxious odours from animal wastes and straw burning do little to improve the public image of agriculture.

The improvement in the production efficiency of intensive agriculture are achieved at a cost of a very large input of energy – the fossil fuel subsidy.

Labour demand in UK agriculture is steadily declining and the holding size is more than double that for any of its European neighbours.

During the early stages of mechanisation, the fall in the size of the total workforce is accompanied by an increase in tractor ownership.

When farms are fully mechanised, further labour shedding is associated with a reduction in tractor ownership but the average tractor power is increased to maintain the total tractor power available.

Machinery management is the study of the selection, operation and replacement of farm machines.

The optimum farm power and machinery complement is achieved through a combination of the maximum tractor power demand for a single operation, the tractor fleet size for simultaneous operations, and the tractor power mix to accommodate varying power requirements of different machines.

2

OPERATOR PRODUCTIVITY

APOSTROPHE – SHOW YOUR METAL

At the heart of the corn mill were the millstones themselves.
These were arranged in pairs, the upper 'runner stone' rotating
above the lower fixed 'bed stone' or 'ligger'. The stones were
usually about four feet in diameter, could weigh upwards of a
ton when new, and functioned most efficiently at a speed of
about 125 rev/min. Sources of suitable material were few and
far between; most highly prized of all were the French 'burrs',
skilfully pieced together from blocks of a particular type of
quartz quarried near Paris. This material only occurred in small
deposits and it was necessary to shape and match the pieces
with great care. They were jointed in cement, bound with iron
hoops and backed with plaster of Paris. Pairs of French 'burrs'
still survive in many mills and are easily identifiable by their
patchwork appearance.

The work surfaces of the stones required careful preparation
and skilled stone dressers were craftsmen in their own right.
When dealing with a new stone, it was necessary first to
produce a perfectly smooth surface. A laminated wooden
straight edge – checked against a proof staff of cast iron from
time to time – was used with 'raddle', a composition of red

oxide, to detect high spots, and these were rubbed down with a fragment of burr stone. The final aim was to achieve a surface very slightly dished, or hollowed, towards the central eye, ready to receive the system of furrows which performed the actual operation of grinding. In common dressing, the surface was marked out into ten equal 'harps' or sectors divided not by true radii, but by lines drawn at a tangent to the eye of the stone. Each 'harp' was subdivided into alternate 'lands' and 'furrows', the latter cut to a depth which varied between one half and three-quarters of an inch and finished with a sharp arris on one side and an even slope up to the 'land' on the other. Finally, a system of fine parallel grooves, the 'stitching' or 'cracking', was added to the surface of each land, a process calling for both skill and precision, experienced men being able to cut as many as 16 cracks to the inch. The dressing was so arranged that when the stones were in their working position, face to face, the furrows crossed at each revolution, cutting the grain with the action of scissor blades. As the stone became dulled with wear, they ceased to grind efficiently and frequent re-dressing was required. This involved the deepening of the 'furrows' and the renewal of the 'stitching' and, with a pair of French burrs, might become necessary every two or three weeks.

Dressing was performed with the 'bill', a cutting tool resembling a double-edged wedge forged from high carbon steel, and with the pointed 'pick'. These tools, mounted in turned wooden handles or 'thrifts', were used with a pecking action to chip away the surface of the stone. A stock of 'bills' was always maintained as one man would blunt three or four in the course of an hour's work, and a grindstone, therefore, formed an essential part of the equipment of every mill. With the decline of the working mill, many millers were forced to undertake the dressing of their own stones but, in the nineteenth century, pairs of itinerant stone dressers made a living by tramping the country from mill to mill. Such men's hands became discoloured by minute particles of steel struck from the 'bills' which lodged under the skin. Before engaging a man whom he did not recognise, the miller would ask him to '*show his metal*', and judge from the state of his hands whether he was indeed an experienced craftsman.

(Source: Reynolds 1970)

2.1 INTRODUCTION

With the changing demands of agriculture, manual labour has been replaced by operator control of machines. For maximum *operator productivity*, therefore, it is not only the actual task which must come under scrutiny but also the operator/machine interface.

From current farm practice, general management guidelines can be obtained on the average labour requirements and seasonality of the labour demand. In order to improve the use of these manpower resources, detailed examination of the individual tasks using *Work Study* techniques can be combined with the most effective allocation of these resources by *Network Analysis* for a series of interconnected tasks.

As machines become larger and more powerful, the *physical stress* which was involved in completing the work manually is substantially decreased but the *total stress* on the operator continues to increase because of the *mental stress* of decision-making. Total stress can be reduced by careful layout of the workspace and by greater automation of the operator's control functions. In addition, reduced exposure to noise and vibration, and a comfortable workspace free from dust and smells improves operator performance and decreases accident susceptibility.

2.2 EFFECTIVE USE OF MANPOWER RESOURCES

2.2.1 Labour requirement

The best guide to the staff requirements for a farm business is past experience from a similar size and type of holding, but this prior knowledge may not always be available. An indication of the manpower requirements can be obtained from average labour data in terms of the *standard man-days* associated with individual enterprises.

- A **standard man-day** is a normal working day of eight hours for one man operating under average conditions, with a currently acceptable degree of mechanisation.

From this definition, it is obvious that the work content of a standard man-day is far from a precise measure because the manpower requirements for an enterprise are closely related to

Table 2.1 Annual labour requirements for individual enterprises

Crops	Labour demand, standard man-day/ha	Livestock	Labour demand, standard man-day/head
Cereals	1.75	Dairy cows: parlour	5
Straw harvesting	0.75	cowshed	7
Potatoes	22	Beef cows	2
Sugar beet	8	Bulls	4
Vining peas	3.5	Barley beef: 0–1 yr	3
Threshed peas	1.75	Other cattle: 2+ yr	2.5
Field beans	1.75	1–2 yr	2
Oilseed rape	1.75	$\frac{1}{2}$–1 yr	1
Herbage seeds	1.75	Calves: 0–$\frac{1}{2}$ yr	
Kale (grazed)	2	multi-suckled	1
Turnips, swedes:		hand-fed	2
harvested	12	Ewes/rams	0.5
folded	6	Other sheep over 6 mnth	0.3
Hay/silage: 1 cut	2.5	Sows	4
2 cuts	4	Boars	2
Grazing only	1	Other pigs over 2 mnth	0.6
Hay for sale	2.5	Laying hens	0.05

(*Source*: Nix, 1986)

the level of mechanisation, one being a partial substitute for the other. The average annual labour figures for some of the most important crop and livestock enterprises are given in Table 2.1 With a high level of mechanisation, the labour involvement may be 25–30 per cent less than average.

Once the total labour requirement has been calculated from these data, it is usual practice to add 15 per cent to allow for unproductive work on farm maintenance and overheads generally.

2.2.2 Labour efficiency index

The total labour requirement which has been estimated for a particular farm may be expressed as a percentage of the number of standard man-days available to give an *index of labour efficiency*:

$$\text{Labour efficiency index, } \% = \frac{\text{Estimated labour requirement, man-day}}{\text{Labour available, man-day}} \times 100$$

. . .[2.1]

In order to evaluate the labour efficiency index, it is considered that the total number of standard man-days available per year is 300 for a stockman and 250 for other farm staff.

If, for example, the estimated total labour requirement (including unproductive work) is 720 standard man-days for a farm with a regular staff of three men, comprising a stockman and two tractor drivers, then:

$$\text{Labour efficiency index} = \frac{720}{(300 + 250 + 250)} \times 100 = \underline{\underline{90\ \%}}$$

Whilst this level of labour efficiency is by no means perfect, it is unlikely to be a major cause for managerial concern.

2.2.3   Seasonal labour demand

Not only is the total labour requirement important but equally the seasonality of the labour demand. Whilst intensive livestock enterprises can be planned to have a constant labour demand and to achieve a uniform production level with different groups of livestock reaching maturity throughout the year, crop production enterprises have peak labour demands during establishment and harvesting operations. The seasonal variation in the labour requirements is given as a percentage of the total number of standard man-days for different enterprises (Table 2.2).

The most immediate solution for meeting these peak labour demands is either to work longer hours with overtime or by turning to optional sources such as hiring casual labour and/or using a contractor. There are, however, alternative steps which could ameliorate the problem if labour is not considered in isolation from other resources. Greater mechanisation could increase the work output per man, though not necessarily the profit margin. Adjusting the balance of enterprises might spread

Table 2.2 Seasonal labour demand as a percentage of the total number of standard man-days required annually

Farm enterprise	Monthly labour requirements, % of total annual demand											
	Jan	Feb	Mar	Apr	May	June	July	Aug	Sept	Oct	Nov	Dec
Winter cereals	—	—	4	8	4	—	—	22	30	25	7	—
Spring cereals	—	4	22	8	4	—	—	37	14	4	5	2
Potatoes – main crop	1	1	7	20	2	4	4	1	10	38	11	1
Potatoes – early	5	16	19	4	3	25	29	—	—	—	3	1
Turnips, swedes	5	5	5	10	7	20	8	—	2	19	20	20
Sugar-beet	1	—	9	5	16	17	2	—	19	25	22	7
Oilseed rape – Spring	—	4	18	16	5	3	—	—	19	10	5	5
Kale – grazed	10	5	5	25	15	10	—	—	—	10	10	10
Kale – cut	10	5	10	20	10	10	—	—	—	10	15	10
Vining peas	—	12	8	5	4	2	30	30	—	3	6	—
Hay + grazings	—	—	3	10	3	55	25	4	—	—	—	—
Silage – 1 cut + grazings	—	—	15	15	35	37	4	4	—	—	—	—
Silage – 2 cuts + grazings	—	—	11	4	25	28	28	4	—	—	—	—
Grazing only	10	3	20	10	15	17	15	15	5	—	—	—
Dairy cows	10	10	10	9	7	6	7	7	7	8	9	10
Other cattle	12	12	11	7	7	5	5	5	5	8	11	12
Sheep	4	8	30	10	6	10	8	4	4	8	4	4
Pigs	9	9	8	8	8	8	8	8	8	8	9	7

(*Source:* Buckett, 1981)

the labour demand over a longer timespan or transfer the peak demand to a better season of the year when a greater number of work days were available. The interaction of these options is examined more fully under work organisation (see section 9.3) but, before considering the interchangeability of the resources, it is important to ensure that the essential work is undertaken as effectively as possible through the application of work study techniques.

2.3 WORK STUDY

Simplifying the job in hand is as old as industry itself. Whilst this attitude towards activities is largely organised common sense, the approach has been formalised into a productivity science called *work study* which relies less on inspiration and more on routine procedures. These routine procedures seek to ensure that nothing is forgotten, nothing is taken for granted and nothing is left to chance.

- **Work study** examines systematically, objectively and critically all the factors which govern the operational efficiency of any specified activity in order to effect improvement.

The technique provides a means of examining routines developed by habit, through faulty decisions or in an inefficient workplace environment; it can result in better methods of work, better layouts or improved procedures.

Work study comprises two distinct yet completely interdependent techniques, *method study* and *work measurement* (Fig. 2.1). The first of these is concerned with the way in which work is done, that is the *method*, and the second with the value of the *work content* of the task itself.

- **Method study** is the systematic recording and critical examination of the factors and resources involved in existing and proposed ways of doing work, as a means of developing and applying easier and more effective methods and reducing costs.
- **Work measurement** is the application of techniques designed to establish the time for a qualified worker to carry out a specified job at a defined level of performance.

Perhaps the most important feature of work study procedures is their adaptability to different tasks, whether performed entirely

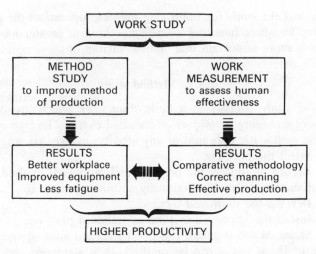

Fig. 2.1 The interdependence of method study and work measurement in work study (adapted from Currie, 1972).

in one location or during continuous movement, whether taking seconds or weeks to complete and whether continuously repetitive or unique. For analytical purposes, the tasks may be graded into five *activity classes* of increasing detail:

○ **process** – sequence of jobs to complete an objective, e.g. *feed processing*;

○ **job** – sequence of operations complete in themselves, e.g. *mixing ration*;

○ **operation** – sequence of work elements as part of a job, e.g. *adding ingredients to mixer*;

○ **work element** – sequence of working movements, readily defined and accurately timed, e.g. *emptying sack*;

○ **therbligs** – elementary divisions of working movements (the term being derived from the name of the originator, Gilbreth, spelt backwards), e.g. *grasp, lift*.

Repetitive farm work as found in the dairy parlour or in the vegetable packing station provides an obvious opportunity for detailed investigation of work elements because even small daily savings accumulate on an annual basis. There is also considerable scope in field work for improving multiple operations in silage making and bale handling where smoothing of seasonal labour peaks is more important than annual savings.

2.3.1 Method study

Method study consists of a basic *framework* which involves a series of *recording techniques* on standard charts. The framework for the systematic appraisal of any activity involves six stages.

(a) **Select** the problem.
(b) **Record** accurately the present (and proposed) method.
(c) **Examine** the method critically and in sequence.
(d) **Develop** the improved method.
(e) **Install** the improved method as standard practice.
(f) **Maintain** the standard practice by regular routine checks.

None of these stages can be omitted, their sequence and their content being important.

(a) and (b) SELECT and RECORD stages

In the *select* stage, the problem is identified and the start and end of the investigation precisely prescribed. The *record* stage which follows is essential to the success of the investigation because the records provide the subsequent basis of both the critical examination and the development of the improved method. The recording techniques vary according to the nature of the activity that is under scrutiny, the most relevant being:

○ **flow process chart**   – sequence of activities of men or materials or machinery;

○ **two-handed process chart** – sequence of activities of an operator's two hands;

○ **multiple activity chart** – sequence of linked activities of a team and/or materials and/or machines on a common timescale;

○ **flow diagram**   – pattern of movement of men, materials or machinery;

○ **string diagram**   – pattern and extent of movement of men or materials or machinery.

(c) EXAMINE stage

Once the existing activites have been investigated using these techniques (see sections 2.3.3 to 2.3.5), all the recorded information is scrutinised and each activity is critically examined to determine whether it may be:

eliminated,

combined with any other part of the job,

changed in sequence to avoid delay, or

simplified to reduce the content of the work involved.

This *examine* stage is expedited by segregating the activities into three groups:

MAKE READY activities;

DO activities;

PUT AWAY activities.

The key 'DO' activities are always examined first because if these can be changed in any way or eliminated, then this will automatically affect the rest.

The critical examination procedure itself is conducted by means of a *questioning sequence*, each activity being subjected to a systematic and progressive series of questions (Table 2.3). The primary questions solicit the facts and the reasons underlying them, but also challenge the existence of an activity, whilst the secondary questions solicit options and opportunities. The questions are asked under five headings which inquire into the purpose of the activity, the place where it is carried out, the sequence of the activity in relation to other activities, the person performing the activity and, finally, the means of completion. Both primary and secondary questions are asked for each aspect before proceeding to the next. Obviously, if the questioning sequence does not establish a purpose for the activity, there is no need to waste further time inquiring into any other aspect of it.

Attitude of mind is vitally important in developing a critical faculty for the examination of the activities and the approach should take the following form:

o **validate** the evidence – examine the facts as they are, excluding assumptions;

o **challenge** everything – avoid adapting the evidence to preconceived solutions;

Table 2.3 Method study – a critical examination sheet for the questioning sequence

Activity aspect	Challenge	Alternative	Opportunity
Purpose	WHAT is achieved?	WHAT ELSE could be done?	WHAT SHOULD be done?
Place	WHY IS IT NECESSARY?	WHERE ELSE could it be done?	WHERE SHOULD it be done?
Sequence	WHERE is it done?	WHEN ELSE could it be done?	WHEN SHOULD it be done?
Person	WHY THERE?	WHO ELSE could do it?	WHO SHOULD do it?
Means	WHEN is it done?	HOW ELSE could it be done?	HOW SHOULD it be done?
	WHY THEN?		
	WHO does it?		
	WHY THAT PERSON?		
	HOW is it done?		
	WHY THAT WAY?		

○ **complete** the full examination – avoid premature conclusions;
○ **justify** the need for an activity before improving the method;
○ **judge** alternatives on facts not suppositions;
○ **learn** from the analysis – involve those associated with the activity.

(d) DEVELOP stage

Appraisal of the existing method provides the guidelines for the *develop* stage, leading to improved methods within a modified workplace environment. Planning a more effective workplace layout requires an understanding of the *principles of motion economy*, namely:
○ **minimum movements;**
○ **simultaneous and symmetrical movements;**
○ **natural movements;**
○ **rhythmical movements;**
○ **habitual movements;**
○ **continuous movements.**
These principles are common to all skilled movements and form the basis on which all movement patterns should be constructed.

Minimum movements
Excessive body movements at the workplace lead to fatigue. Micromotion studies have allocated upper limb movements of increasing severity into five classes:
I fingers only;
II fingers and wrist;
III fingers, wrist and lower arm;
IV fingers, wrist, lower and upper arm;
V fingers, wrist, lower and upper arm, shoulder.
Movement is kept to a minimum by operating within the *normal working area* but is acceptable within the *maximum working area*. For a seated operator, the normal working area in the horizontal plane is bounded by arcs drawn by the right and left hands with the lower arms pivoting about the elbows which are held naturally by the side of the body, i.e. Class III movements (Fig. 2.2). The equivalent maximum working area is bounded by arcs drawn by the right and left hands, with the arms pivoting

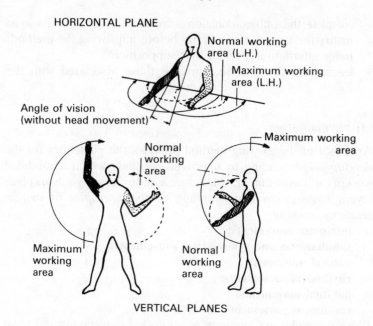

HORIZONTAL PLANE

Normal working area (L.H.)

Maximum working area (L.H.)

Angle of vision (without head movement)

Normal working area

Maximum working area

Maximum working area

Normal working area

Normal working area

VERTICAL PLANES

Fig. 2.2 Normal and maximum working areas in both the horizontal and vertical planes (after Currie, 1972).

from the shoulder. Objects placed between the normal and the maximum working areas can be reached by a Class IV movement. Anything placed outside the maximum working area can only be reached by a Class V movement involving body stretching or bending. There are corresponding normal and maximum working areas in the vertical planes (Fig. 2.2).

Simultaneous and symmetrical movements
The anatomy of human beings is such that balanced movements are less tiring than unbalanced ones. Walking with a heavy bucket of water in one hand requires extra, unproductive, muscular exertion for counterbalancing, whereas half the load in each hand restores bodily equilibrium. It is also less tiring when both hands undertake tasks which are mirror images of each other than it is to keep one hand inactive.

Natural movements
Correct use of muscles and posture in weight lifting, for example, improves productivity as well as reducing accidents.

Rhythmical movements
Establishing a rhythm leads to faster movements with less fatigue. Hesitations are eliminated by the regular repetition of a movement pattern, often characterised by the accentuation of part of the cycle. The provision of music, or chanting, to maintain a suitable rhythm is a well-known practice.

Habitual movements
With our natural ability to form habits, a set of movements performed habitually become almost a reflex action, involving little mental effort. For example, commodity supplies which are always located in the same position on a vegetable packing line can be picked up without searching or fumbling and without excessive eye movement. It is equally important to remember that incorrect movements can also become habitual – and bad habits are difficult to eradicate.

Continuous movements
Smooth, curved and continuous movements are preferable to straight movements involving sharp changes of direction. Frequent abrupt movements induce fatigue because they require a constant state of tension in the muscles. A uniform pace is preferred to either speeding up or slowing down movements.

During the develop stage, work measurement (see section 2.3.2) may be required to assess the economic manning levels for a job. It is only by this means that slack time can be removed from the method.

(d) and (e) INSTALL and MAINTAIN stages

The *install* stage is when an agreed method is ready to be put into practice. The success of the whole investigation depends on the active support of everyone concerned and adequate planning for the change from the existing method to the proposed method, including a rehearsal of the agreed method.

After the improved method has been installed and is operating satisfactorily, the benefits may not persist without additional effort. The *maintain* stage involves a regular review of current practices to reveal any changes of method which gradually occur either through minor innovation or through a gradual drift towards unacceptable shortcuts.

2.3.2　Work measurement

As labour costs represent such a large proportion of total production costs, inaccurate estimates of work times present a major management problem. The actual time taken to complete a job depends on the rate of working maintained by a particular individual. Variations in this rate of working arise from a variety of different causes of which some may be the responsibility of management, such as waiting for work; some may be the responsibility of the operator, such as lengthy chats to a team mate between cyclic operations; and some may be contingencies, such as consultation with the farm grieve or foreman.

Work measurement aims to identify *standard rates of working* for planning purposes.

- The **standard performance** is the optimum rate of output that can be achieved by a qualified operator at the desired level of effectiveness and when the appropriate relaxation allowance is taken in full.

This standard performance can be maintained as an average for the working day and throughout the week without over-exertion and consequent detriment to health.

The practice of *work measurement* involves four essential steps.

(a)　**Divide job into work elements**. These elements are usually less than 60 seconds duration and are chosen to separate light work from heavy work. They are separated, where possible, by audible break-points. For example, the element for unhitching a trailer is completed when the trailer hits the ground.

(b)　**Record basic times using time study**. Repetitive work elements are directly observed while in progress, the accuracy of results being improved as the number of observations increases.

(c) **Assess the relaxation allowance**. The proportion of rest-time varies with the mental and physical effort involved in each work element. Light, medium and heavy jobs have relaxation allowances of 10, 20 and 30 per cent of the effort portions of the element, respectively.

(d) **Obtain standard performance times**. The times for the effort portion and the relaxation allowances are combined.

Adequate detail must be recorded in the time study for effective analysis. By way of illustration, consider the time study of turn round times, averaged from seven observations, for a trailer at a silage clamp. A general study of the system, recording few details, is simply a statement of current practice and nothing more can be derived from it.

General time study of system (low detail)

Element break-points	Recorded time, min : s	Element	Lapsed time, min : s
Stop at clamp	0 : 00		
Go	4 : 00	Turn round	4 : 00

In constrast, a more detailed study of the routine may identify a number of opportunities for improvement.

Detail time study of the tip routine

Element break-points	Recorded time, min : s	Element	Lapsed time, min : s
Stop at clamp	0 : 00		
Reverse and *stop*	0 : 30	Reverse	0 : 30
Off tractor, walk to *back*	0 : 50		
Door *off*	1 : 50	Mount/dismount	1 : 20
On tractor, *start* tip	2 : 10		
Load *out*	2 : 40	Tip	0 : 30
Off tractor, walk to *back*	3 : 00		
Door *on*	3 : 40	Tailgate off/on	1 : 40
On tractor and *go*	4 : 00	TOTAL	4 : 00

Four elements are now identified in place of one element for the general study. The most time is spent getting on and off the tractor and in removing the tailgate. Taking this information forward to an investigation by method study, two options emerge.

Method study – system alternatives

(i) Helper takes off tailgate
(no dismount)

Element	Lapsed time min : s
Reverse	0 : 30
Door off/on	1 : 40
Tip	0 : 30
TOTAL	2 : 40

(ii) Automatic tailgate
(no dismount)
(no tailgate off/on)

Element	Lapsed time min : s
Reverse	0 : 30
Tip	0 : 30
TOTAL	1 : 00

Thus, it is perfectly feasible to install a tip routine which is four times faster than the present system by investing in automatic tailgates for the trailers (Fig. 2.3).

The availability of task times helps designers compare various systems with particular emphasis on labour economy. Although such data for agricultural operations are fairly sparse, there are

Fig. 2.3 An automatic tailgate on a trailer tipping grass at a clamp silo.

important exceptions, such as the check list for the labour
requirements of beef housing prepared from a Scottish study of
commercial beef units for not less than 125 animals (Tables 2.4
and 2.5). The labour demand is governed by the extent of
mechanisation, with manual systems having not only a higher
time requirement but also a higher level of physical effort as well.
When designing new systems, care must be taken to eliminate

Table 2.4   Task times for feeding silage, turnip, concentrates and hay in
commercial beef units

Feeding system	Daily time per animal, min
Mobile equipment	
Silage delivered on buckrake from silo, concentrates manually spread	0.35
Silage delivered in forage wagon loaded with industrial fore-end loader, turnips in converted muck spreader, concentrates distributed from mobile hopper	0.29
Silage delivered directly into troughs using tractor and fore-end loader, concentrates manually distributed	0.50
Silage, turnips and concentrates all delivered in converted muck spreader, all loaded with tractor and fore-end loader	0.64
Zero grazing	
Grass cut with forage harvester, and delivered in self-emptying forage box, concentrates manually distributed	0.45
Fixed equipment	
Tower silo, top unloader and conveyor belt into trough	0.30
Conveyor belt and loading platform, with silage, turnips and concentrates manually loaded onto the conveyor belt and transported into troughs	0.65
Traditional manual system	
Cutting turnips and delivering by wheelbarrow to trough, and manual feeding of concentrates and hay	1.20
Self-feed silage	
Maintenance of feeding area, moving feed barrier closer to silage face, tractor scraping slurry from the feeding area	0.30
Distribution of concentrates	
Spreading concentrates from 50 kg sack into trough, including carrying the sack over 15 m	0.35
Spreading 100 kg of concentrates from a trolley using a scoop	8.40

(*Source*: Cermak and Ross, 1977)

Table 2.5 Task times for bedding and effluent disposal in commercial beef units

System	Daily time per animal, min
Bedding operation	
Courts – spreading 3 kg of straw per animal using bales	0.35
Cubicles – cleaning bed and spreading 0.5 kg of sawdust	0.03
Farmyard manure disposal	
Handling farm yard manure from court using fore-end loader and spreading with muck spreader on field 0.8 km distant	0.33
All as above except for handling by industrial loader	0.04
Slurry disposal	
Emptying slatted slurry cellar with 3 months' storage capacity and spreading with vacuum tanker (5000–8000 litre) on a field 0.8 km distant	0.05

(*Source*: Cermak and Ross, 1977)

those areas where deliveries of feed have to be manually handled.

2.3.3 Method study recording techniques – process charts

Any task may be divided into five *types of activity* represented by the following symbols:

Symbol	Activity	Effect
○	OPERATION	Produces, accomplishes, furthers the process
⊅	TRANSPORT	Travels
▽	STORAGE	Holds, keeps or retains
D	DELAY	Interferes or delays
□	INSPECTION	Verifies quantity and/or quality

In a flow process chart, these symbols are used to represent the steps in a procedure whereas, in a two-handed process chart, they represent the elements of the work cycle. The activities may relate either to the operator or to the product but not to both in the same chart.

The process of bedding a cattle yard is used as an example of the application of a *flow process chart*, using a standard format

(Fig. 2.4). Each activity detail is specified, the type of activity identified, and the time recorded, as well as notes of distances travelled and any relevant comments. From this record, the number of each type of activity and their total lapsed time can be collated in the summary table. Even without detailed scrutiny

PROCESS CHART, MAN/PRODUCT ANALYSIS

PRESENT/PROPOSED METHOD   SUMMARY

PROCESS : *Bedding cattle yard*

CHART BEGINS : *Man at barn*

CHART ENDS : *Man proceeds to next job*

CHARTED BY : *John Smith*

AT : *Hilltop Farm*

DATE : *20.5.85*

OPERATOR(S) : *A. Bloggs*

ACTIVITY	PRESENT No	PRESENT Time	PROPOSED No	PROPOSED Time	DIFFERENCE No	DIFFERENCE Time
○ OPERATIONS	10	4:01				
☐ INSPECTIONS						
▷ TRANSPORTS	8	3:15				
▽ STORAGES						
D DELAYS						
TOTAL						
DISTANCE TRAVELLED	71 m			m		m

DETAIL	MOVEMENT CLASS	OPERATION / INSPECTION / TRANSPORT / STORAGE / DELAY	DISTANCE TRAVELLED m	ELAPSED TIME min:s	RECORDED TIME min:s	NOTES
Picks up first bale		①☐▷▽D		0:10	0:10	
Carries bale to yard		○☐②▽D		1:08	0:58	
Throws bale over yard wall		②☐▷▽D		1:21	0:13	
Returns to barn		○②▷▽D		1:56	0:35	
Picks up second bale		③☐▷▽D		2:43	0:47	Wedged in stack
Carries bale to yard		○③▷▽D		3:26	0:43	
Throws bale over yard wall		④☐▷▽D		3:38	0:12	
To gate of yard		○④▷▽D		3:43	0:05	
Opens and closes gate		⑤☐▷▽D		3:55	0:12	
Proceeds into yard		○⑤▷▽D		4:05	0:10	
Unties first bale		⑥☐▷▽D		4:15	0:10	
To second bale		○⑥▷▽D		4:23	0:08	
Unties second bale		⑦☐▷▽D		4:33	0:10	
Spreads second bale		⑧☐▷▽D		5:08	0:35	
To first bale		○⑦▷▽D		5:32	0:24	
Spreads first bale		⑨☐▷▽D		6:37	1:05	
Proceeds out of yard		○⑧▷▽D		6:49	0:12	
Opens and closes gate		⑩☐▷▽D		7:16	0:27	
		○☐▷▽D				
To next job		○⑨▷▽D				
		○☐▷▽D				
		○☐▷▽D				
		○☐▷▽D				
		○☐▷▽D				
		○☐▷▽D				

Fig. 2.4 Standard format for a flow process chart.

of the record, it is almost self evident that immediate improvements could be introduced by transporting two bales in one trip instead of individually. Alternatively, a radical change to large roll bales, mechanically transported less frequently, would

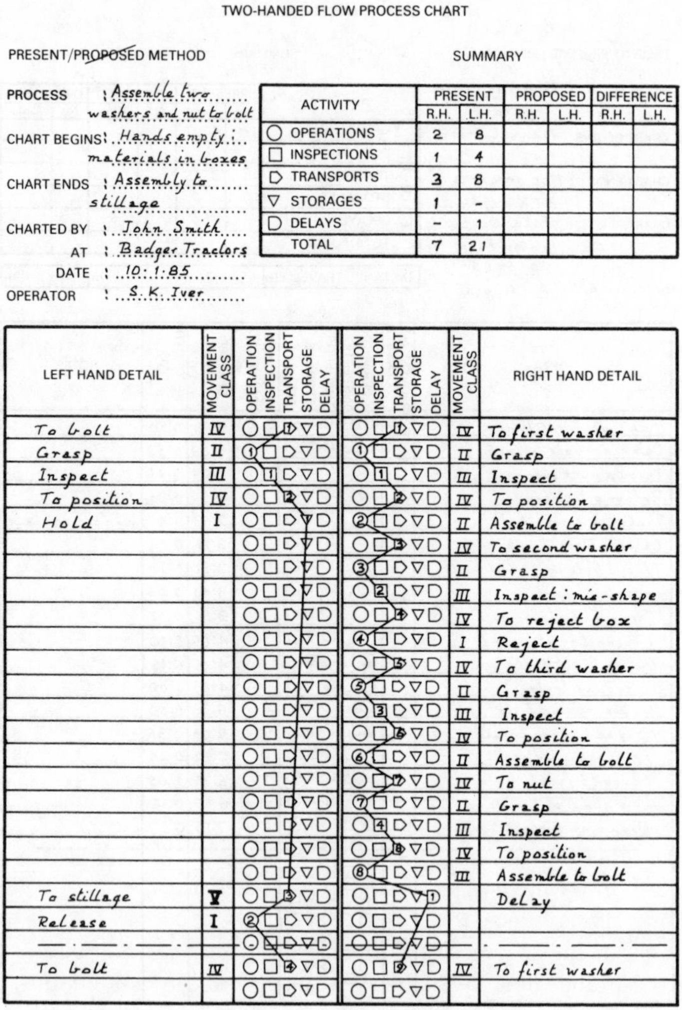

TWO-HANDED FLOW PROCESS CHART

PRESENT/PROPOSED METHOD

PROCESS : *Assemble two washers and nut to bolt*

CHART BEGINS : *Hands empty, materials in boxes*

CHART ENDS : *Assembly to stillage*

CHARTED BY : *John Smith*

AT : *Badger Tractors*

DATE : *10:1:85*

OPERATOR : *S.K. Iver*

SUMMARY

ACTIVITY		PRESENT		PROPOSED		DIFFERENCE	
		R.H.	L.H.	R.H.	L.H.	R.H.	L.H.
○	OPERATIONS	2	8				
□	INSPECTIONS	1	4				
▷	TRANSPORTS	3	8				
▽	STORAGES	1	–				
D	DELAYS	–	1				
	TOTAL	7	21				

LEFT HAND DETAIL	MOVEMENT CLASS	OPERATION INSPECTION TRANSPORT STORAGE DELAY	OPERATION INSPECTION TRANSPORT STORAGE DELAY	MOVEMENT CLASS	RIGHT HAND DETAIL
To bolt	IV	○□▷▽D	○□▷▽D	IV	To first washer
Grasp	II	○□▷▽D	○□▷▽D	II	Grasp
Inspect	III	○□▷▽D	○□▷▽D	III	Inspect
To position	IV	○□▷▽D	○□▷▽D	IV	To position
Hold	I	○□▷▽D	○□▷▽D	II	Assemble to bolt
		○□▷▽D	○□▷▽D	IV	To second washer
		○□▷▽D	○□▷▽D	II	Grasp
		○□▷▽D	○□▷▽D	III	Inspect : mis-shape
		○□▷▽D	○□▷▽D	IV	To reject box
		○□▷▽D	○□▷▽D	I	Reject
		○□▷▽D	○□▷▽D	IV	To third washer
		○□▷▽D	○□▷▽D	II	Grasp
		○□▷▽D	○□▷▽D	III	Inspect
		○□▷▽D	○□▷▽D	IV	To position
		○□▷▽D	○□▷▽D	II	Assemble to bolt
		○□▷▽D	○□▷▽D	IV	To nut
		○□▷▽D	○□▷▽D	II	Grasp
		○□▷▽D	○□▷▽D	III	Inspect
		○□▷▽D	○□▷▽D	IV	To position
		○□▷▽D	○□▷▽D	III	Assemble to bolt
To stillage	V	○□▷▽D	○□▷▽D		Delay
Release	I	○□▷▽D	○□▷▽D		
— · — · — · —		○□▷▽D	○□▷▽D		· — · — · — · —
To bolt	IV	○□▷▽D	○□▷▽D	IV	To first washer
		○□▷▽D	○□▷▽D		

Fig. 2.5 Standard format for a two-handed flow process chart.

produce more dramatic improvements but would also involve investment appraisal as well as method study.

The *two-handed process chart* provides an overall picture of activities at a work-place (Fig. 2.5). When assembling two washers and a nut to a bolt, only the first activity is carried out by both hands simultaneously. Thereafter, only one hand is carrying out the activity. For this detail of method study, it is sometimes helpful to identify, in separate columns on the chart, the movement class (see section 2.3.1) for each activity.

The use of a two-handed process chart is appropriate to the design layout for a potato grader where diseased or damaged produce is manually removed. The crop is conveyed on rotating rollers so that all parts of the tubers can be inspected as they pass the operator. Grading efficiency, as measured by the percentage of the defective tubers removed, is improved from 60 to 80 per cent by having two operators opposite each other so that both ends of each potato can be inspected as it rotates. An interesting aspect of design, however, is that contra rotation of the potatoes relative to the direction of travel must be avoided because the visual effect can lead to the phenomenon called 'belt sickness'. The optimum width of the inspection table and the speed of conveying is dictated by the need to complete full inspection within the normal working area (see section 2.3.1). At a conveyor speed of 0.1 m/s and a roller speed of 40 rev/min, there are two revolutions of the potato within a 300 mm translation. Under these conditions, one operator can inspect between 250 and 300 potatoes per minute, this inspection rate being more significant than the proportion of objects rejected.

For each defective tuber, the hand movements for removal are:
(a) move hand and arm from reject chute to next defective potato,
(b) grasp potato,
(c) transport to reject chute,
(d) release.
These movements are repeated regularly using one or both hands, as necessary.

One of the recent advances in potato grading technology is to increase the time available for inspection by eliminating the transport phase of the grading process (Fig. 2.6). The position of the defective tuber can be automatically identified when

(a)

(b)

Fig. 2.6 Potato grading using a hand-held electronic sensor to identify defective tubers which are later automatically rejected at the end of the conveyor: (a) hand-held sensor; (b) rejection flaps (from SIAE).

touched by a hand-held electronic sensor or 'magic wand' which then emits a vertically polarised radio frequency signal. The signal is received by a matrix of coils under the roller deck and synchronised with the roller conveyor motion to operate solenoid-actuated rejection fingers when the potato reaches the end of the conveyor. Thus, operator activity is restricted to moving from one defective potato to the next and output is increased by 50 per cent. Operator alertness is also increased because there is greater satisfaction derived from 'hitting' a defective tuber than in grasping it, particularly if the potato is visually unattractive because of disease or if it is physically unpleasant to the touch because of decay.

Despite these advances in automatic rejection, the minimum time per object for the inspection activity is dictated in large measure by crop quality; the greater the range of defects being assessed, the longer the inspection time. In a comparison of two samples, both containing 30 per cent of the objects defective, the grading efficiency was 3 per cent lower for the sample with a mixture of two defects than for the sample with a single defect. This lower efficiency indicates that differentiating between multiple defects causes mental confusion, an ergonomic aspect which is more fully discussed in section 2.5.1.

2.3.4 Method study recording techniques: multiple activity charts

Multiple activity charts may be used when it is necessary to consider team work in which several distinct operations are involved, such as silage making (Fig. 2.7). Here, the regular cycles of operations are:
1. forage harvesting to load an in-line trailer;
2. transporting forage from field to silo;
3. spreading the forage to fill the silo evenly.
The forage harvesting and transporting cycles are linked at the changeover of full and empty trailers which involves a hitching and unhitching activity. The transporting and spreading cycles are linked at the trailer unloading activity which interrupts spreading. From the multiple activity chart (Fig. 2.8a), it is evident that transport is adequate because both the forage harvesting cycle and the spreading cycle include regular delays. As the waiting

Fig. 2.7 The silage-making sequence showing: (a) forage harvesting with rear delivery to an in-line trailer; (b) changing trailers; (c) transport; and (d) filling a clamp silo using a push-off buckrake.

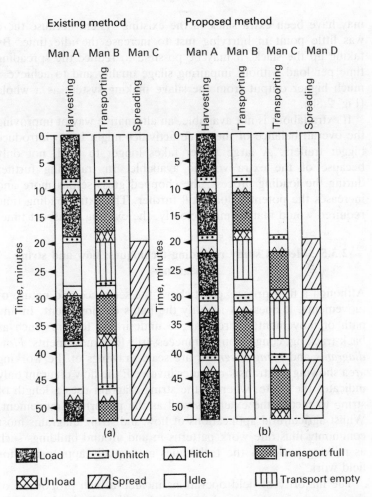

Fig. 2.8 Multiple activity charts for silage making: (a) before method study; (b) a possible alternative with one extra man.

time for the forage harvester is almost equivalent to the time required to load a trailer, the proposed method requires a second transport tractor and driver, plus an extra trailer. This eliminates the harvesting delays but transfers the critical operation from the transport cycle to the spreading cycle. According to the original activity chart there is insufficient waiting time at the silo to accommodate twice the number of trailer loads. However, there may have been extra slack in the existing system because there was little point in hurrying just to increase the idle time. By taking up the slack, it may be possible to reduce the spreading time per load without impairing silage quality and to achieve a much higher output from the silage making system as a whole (Fig. 2.8b).

If extra labour is not available, an alternative way of improving the overall balance of the three activities might be to introduce bigger trailers. A large trailer takes longer to load, not only because of the extra volume available. In travelling further during the loading process, the chopped grass settles more and increases the potential load still further. The extra loading time required would match, more closely, the existing transport time.

2.3.5 Method study recording techniques: flow and string diagrams

Although a flow process chart shows the sequence and nature of movements, it does not identify the paths of movement. In any path of movement, there may be undesirable features such as back-tracking, congestion or unnecessarily long movements. *Flow diagrams* and *string diagrams* are scale drawings of the working area showing the transport routes involved; the flow diagram only indicates the route, whereas the string diagram uses a length of string to record the extent as well as the pattern of movement. Whilst agricultural applications of flow and string diagrams most commonly illustrate work patterns in and around buildings such as milking parlours, the techniques are equally appropriate for field work.

The majority of field operations include a high proportion of time on transportation and materials handling. Reduction of idle running can achieve substantial savings, not only in time but also in soil damage. For example, a common method of slurry spreading is shown in Fig. 2.9(a) where coverage of a square field commences from one of the far sides and gradually proceeds closer to the gate. At the start of the operation, the full slurry tanker tracks the complete length of one headland and returns empty across the diagonal of the field. A more effective method is shown in Fig. 2.9(b) where the field is divided into blocks. The size of these blocks need not be equal and can be adjusted to suit

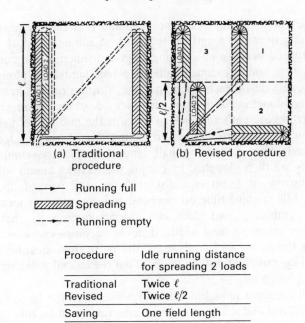

(a) Traditional procedure   (b) Revised procedure

— Running full
▨▨▨▨ Spreading
- - - Running empty

Procedure	Idle running distance for spreading 2 loads
Traditional	Twice ℓ
Revised	Twice $\ell/2$
Saving	One field length

Fig. 2.9 Alternative methods of spreading slurry with a trailed tanker in the field, showing the opportunity for major saving in idle running distances (after Culpin, 1975).

the distance covered by the first pass of the machine. In order to simplify the diagram, however, the idle running time with the improved method is shown to be half that of the original method. The amount and concentration of tracking is also reduced in the improved method. The obvious advantage demonstrated by the use of a flow diagram makes further recording with a string diagram irrelevant in this case.

Preparing a field for conventional ploughing is a more complex problem which lends itself not only to method study and the recording of route distances on a string diagram but also to subsequent mathematical analysis to identify the optimum operational procedures (see also section 3.8.3 and Appendix A1). Consider a 12 ha field, approximately rectangular, which requires headlands and sidelands round the perimeter because the sides of the field are not quite parallel or regular. The field also must be divided into sections or *lands* of convenient size to minimise the idle running time along the headland between each bout of ploughing over the length of the field. A minimum of $1\frac{1}{2}$ lands are required because an opening ridge is formed at the mid-line of the first and subsequent alternate half lands. By ploughing along one side of each opening ridge, turning on the headland and then ploughing back along the other side of the opening ridge, the furrows are gathered or lean towards the mid-line of the half land until the quarter lands on both sides of every ridge are ploughed. The intervening half lands between the completed sections are then ploughed by casting the furrows to end with an open furrow or finish equi-distant from each pair of opening ridges. Idle running time on the headland progressively increases during gathering and decreases during casting so that, by judicious choice of land width, it never becomes excessive. The greater the number of lands used, the more time must be spent in marking out and finishing the field but the less idle running time during ploughing.

Traditionally a ploughman would walk round the field, pacing the headland and sideland width at intervals, perhaps three times along each side of the field, and digging a divot of turf to act as a guide marker when using the plough to score a shallow furrow as a permanent indication of headland width (Fig. 2.10(a)). This is a time-consuming activity and even a brisk walk round the perimeter of the field would take twenty minutes. It would be far better to drive round in the tractor, stopping briefly to pace the headland width as required. Once the first, transport circuit of the field is completed, scoring the shallow furrow is carried out during a second circuit (Fig. 2.10(b)).

The next stage is to use the tractor again for transport along-side the field boundary chosen as the base line for setting up the opening ridges in parallel. Again, stopping briefly three times to pace the distance accurately to the first opening ridge and setting up a ranging pole on each occasion, the tractor is driven to the far end of the field and lined up with the ranging poles so that a shallow furrow can be drawn on the return pass towards the gate end of the field. Before moving away, however, the distance is paced out to the next opening ridge and a fourth ranging pole set up to provide the longest sight line for accurately scoring the next opening ridge (Fig. 2.10(c)). During the return pass to the gate end of the field, the tractor is stopped at each ranging pole

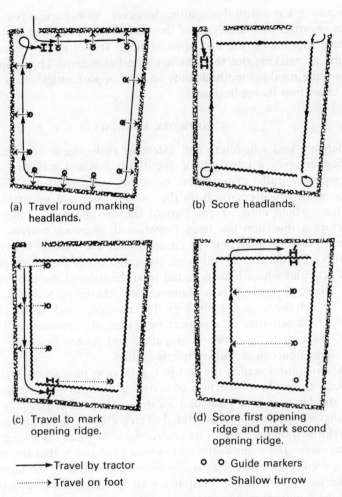

(a) Travel round marking headlands.

(b) Score headlands.

(c) Travel to mark opening ridge.

(d) Score first opening ridge and mark second opening ridge.

──▶ Travel by tractor   O O Guide markers

┄┄▶ Travel on foot   〰〰〰 Shallow furrow

Fig. 2.10 A sequence of string diagrams being used to record the operator's travel movements both by tractor and on foot, when preparing a field layout for conventional ploughing.

which is moved across to the next opening ridge. By the time the first ridge is scored, the second opening ridge is marked with ranging poles and the last ranging pole to be removed from the first ridge is transferred to the third opening ridge (Fig. 2.10(d)). In this way, wasted time is reduced and no other assistance is

required. Even with this routine, however, further scope remains
for improvement. One pass of the field could be eliminated if the
ranging poles for the first opening ridge are set up at the same
time as marking out the headlands and sidelands. This emphasises the need for method study and the opportunities which can
accrue from its application.

2.4 NETWORK ANALYSIS

Planning and scheduling and extensive realignment of a major
road involves a labyrinth of apparently isolated activities and
temporary traffic management measures as different sections of
the work which encroach on the existing lanes are undertaken.
Chaos would ensue if any part of the existing road is dug up
before a diversion has been constructed. *Network analysis* not
only provides a method of ensuring that the various activities
proceed in the correct order but also can identify the critical path
of activities which influences the total duration of the project.

- **Network analysis** is a planning and scheduling technique in
  which the sequence, the logical relationships and the duration
  of all activities in a project or system are represented in a
  diagram by a network of paths, with events marking the
  commencement and completion times.

In agricultural terms, the 'project' is a crop production system
which is simply a collection of activities necessary to produce a
particular crop in a particular location. These activities may be
either a biological process related to crop growth, or an operation
performed on the crop or its environment to guide the biological
processes. The main distinction between the two is that the time
required for a biological process is not controlled by the farmer,
whereas (within limits) the time for an operation can be adjusted
to meet management constraints.

2.4.1 Construction of the network

The network diagram consists of a number of *activities*, each
represented by an arrow (Fig. 2.11). Each arrow is separated by
an *event* or node, represented by a circle which contains a
reference number so that it can be readily identified.

When constructing the network it is necessary to consider each
activity in turn and ask:

(a) what activity precedes it?

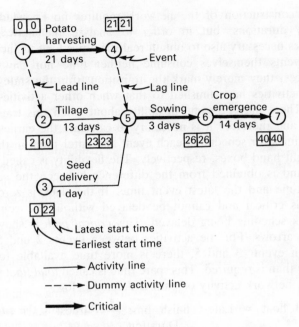

Fig. 2.11 A network diagram with activities separated by events, each with the earliest and latest time.

(b) what activities run concurrently?
(c) what activity follows it?
(d) what path of activities control the start time?
(e) what path of activities control the finish time?

Thus for establishing a crop of winter wheat after potatoes, tillage cannot begin until harvesting of the previous crop is underway but can take place in parallel thereafter, provided that tillage proceeds at the same rate or slower than potato harvesting. This interaction is shown by the *dummy* activities which are represented by dotted arrows, the slope indicating *lead lines* at the start of each activity and *lag lines* at the finish of each activity running concurrently to produce *laddering*. Sowing could be shown as a 'third rung in this ladder' but, in the example, it is considered that no further operators are available. There can be no dispute, however, that sowing cannot start before the seed is delivered and that crop emergence is some time after the completion of sowing unless excessive weather delays have occurred during sowing.

The construction of the network requires no knowledge of activity durations; but in order to analyse the network, it becomes necessary also to obtain realistic times for each activity. The events themselves consume neither time nor any other resources, they merely mark the transition from the state when some activities have finished to that when other activities may start. There may be some flexibility about when this transition occurs and therefore it is necessary to calculate the earliest and latest times for scheduling each event and enter them in the left and right hand boxes, respectively. The flexibility is called *event slack* and is obtained from the difference between the earliest event time and the latest event time. If the slack is zero, the event is critical and cannot be delayed without the complete network schedule being delayed. This *critical path* is shown by double arrows. For the activities between events 2 and 5 and between events 3 and 5, there is more time available for the activity than is required. This spare time is called *total float* which for any network activity or path is defined below.

$$\text{Total float} = \text{Latest finish time} - \text{Earliest start time} - \text{Duration}$$

The seed delivery activity has the greatest slack and therefore allows considerable flexibility in scheduling farm transport when perhaps weather has delayed field work.

2.4.2 Resource allocation

Work organisation is seldom dependent on time analysis alone and involves allocation techniques for other resources such as labour, equipment, material, space or finance. The simplest technique for this purpose is called *resource aggregation*. The resources required by each activity are accumulated for each resource category against a time scale without any consideration of the availability of the resources. In the simple network in Fig. 2.12, the numbers above the activity arrows are the resource units required, whilst the dates are shown for each event. The resource accumulation chart in Fig. 2.13(a) is based on each activity commencing at its earliest start time. The resource requirements are extremely variable so that some reorganisation is desirable, unless casual labour is readily available. Improve-

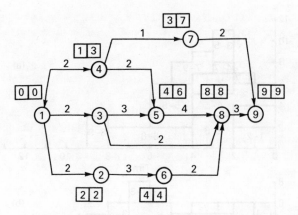

Fig. 2.12 .A simple network with resource requirements for each activity as well as a time schedule (after McLeod and Staffurth, 1968).

ment in resource allocation can be achieved in two ways, depending on whether or not the duration of network must be kept to a minimum. If the minimum duration must be maintained, then it may be possible by the use of float to achieve some *resource smoothing*, that is, eliminating the worst variations in resource requirements (Fig. 2.13(b)). When some relaxation of time schedule is acceptable, *resource levelling* can be used to ensure that project duration is minimised, subject to the constraint that a predetermined resource level is never exceeded (Fig. 2.13(c)).

Whilst it is possible to achieve the best solution by trial and error for only one type of resource in a small network, multiple interactions become extremely complex. In addition, cumulative process duration probabilities must be included in agricultural systems to account for weather-dependent activities. The computer methodology is still under development but manual applications of resource allocation techniques are further discussed in section 9.3.2.

2.5 ERGONOMICS

In agricultural mechanisation just as in any other industrialisation process, the advent of the machine has introduced not only major

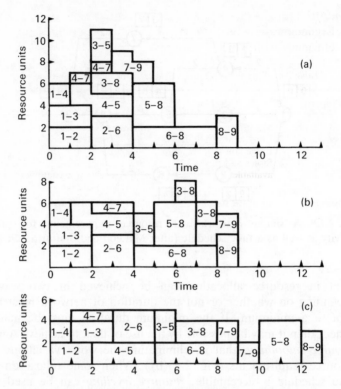

Fig. 2.13 Resource allocation charts from the network in Fig. 2.12 showing alternative organisational strategies: (a) resource accumulation; (b) resource smoothing; (c) resource levelling (after McLeod and Staffurth, 1968).

changes in work organisation but also gradual alterations to the acceptable working conditions. The machine modifies the task that man must perform; it makes new and different demands on their resources and energies; and it replaces manual labour with operator skills. These operator skills are far more valuable than manual labour because productivity is enhanced by machines of ever increasing output. The greater earning potential of operator skills, the greater the financial flexibility which can be used to satisfy the expectations for more leisure time by adjusting the work patterns. This, in turn, increases the demand for higher productivity during the work cycle, and the need to apply *ergon-*

omics to improve the design of the workplace environment.

- **Ergonomics** is the study of the factors influencing the efficiency with which a man may carry out his work.

Table 2.6 Fitts list – relative advantages of men and machines

Property	Machine	Man
Speed	Much superior.	Lag one second.
Power	Consistent at any level. Large constant standard forces and power available	1.5 kilowatts for about ten seconds. 0.4 kilowatts for a few minutes. 0.1 kilowatts for continuous work over a day
Consistency	Ideal for routine repetition precision	Not reliable – should be monitored. Subject to learning and fatigue
Complex activities	Multi-channel	Single channel. Low information throughput
Memory	Best for literal reproduction and short-term storage	Large store multiple access. Better for principles and strategies
Reasoning	Good, deductive. Tedious to reprogramme	Good, inductive. Easy to reprogramme
Computation	Fast, accurate. Poor at error correction	Slow. Subject to error. Good at error correction
Input	Some outside human sense, eg radioactivity	Wide range (10^{12}) and variety of stimuli dealt with by one unit, eg eye deals with relative location, movement and colour
	Insensitive to extraneous stimuli. Poor pattern detection	Affected by heat, cold, noise and vibration. Good pattern detection. Can detect very low signals. Can detect signal in high noise levels
Overload reliability	Sudden breakdown	Graceful degradation
Intelligence	None. Incapable of goal switching or strategy switching without direction	Can deal with unpredicted and unpredictable. Can anticipate. Can adapt
Manipulative abilities	Specific	Great versatility and mobility

(*Source*: Singleton, 1974)

The aim of the designer is to use the machine to enchance and extend the capabilities of the operator ráther than to replace him. In the list of the relative advantages of men and machines (Table 2.6), the superiority of the operator for controlling machine output is clearly identified but, equally, his unreliability and his vulnerability to environmental hazards is also underlined.

An operator interacts with a machine in two ways: first, the operator receives and processes information derived from the machine display panel and from the machine output; and secondly, the operator applies that information by actuating controls to adjust the performance of the machine. The reception of the information is by means of the operator's sensory processes (i.e. visual, auditory and tactile), whilst the output elements are the limbs for activating the controls (Fig. 2.14). Linking the displayed input with the output control actions is the operator's capacity to process information and hence perform the task – his level of skill.

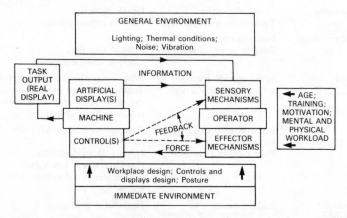

Fig. 2.14 Operator/machine interaction together with factors affecting performance.

In exercising his skill, an operator of a farm machine involves a variety of psychological processes which define his total performance (Table 2.7). Some of these functions, of course, can be incorporated into the machine, depending on the complexity

Table 2.7 Psychological processes which define operator performance

Process	Characteristics
Identification	The perceiving of objects and cues in his operating environment and understanding their relevance to his current task.
Discrimination	Identifying the elements and cues in his current task and perceiving the differences between them.
Scaling	Assigning values to the information with which he is presented: scaling may be simple ordinal (yes/no) or may be finely structured.
Integration	Assimilating information from a variety of sources and tying them together into some meaningful whole which is relevant to the current phase of the task.
Remembering	Using part information which he has stored in some form either from the immediate past or from previous experience of the same task.
Judgement	Weighing the different information and combining it.
Decision	Combination of judgements and commissioning a course of action.
Ordering	Arranging the temporal order in which actions are to be commissioned.
Timing and Co-ordination	Determining the exact points in time when actions should be initiated or terminated.
Monitoring	Collecting information to confirm that the actions and responses are as planned.
Anticipation	Planning future actions.

(*Source*: Easterby, 1975)

of the technology which is economically justified. Nonetheless, the skill of the operator is the way in which he combines *all* the elements into an integrated activity.

2.5.1 Information reception and processing

Modern concepts of information theory, developed for the computer, can be applied to the human brain as it behaves in a similar way to electronic single channel mechanisms. The measure of information gain is called the *bit* which is a contraction of 'binary digit'.

- A **bit** is the amount of information which is contained in a single choice between two alternatives.

For a number of alternatives, N_a, which are all equally probable in a given situation, the mathematical definition of a bit is:

$$H_i = \log_2 N_a \qquad \qquad ...[2.2]$$

Information, bit = $\log_2$ (Number of alternatives)

If the number of alternatives in eqn [2.2] are not all equally probable, then:

$$H_i = \log_2(^1/_p) \qquad \qquad ...[2.2a]$$

Information, bit = $\log_2$ (1/Probability of event occurring)

The rate of information handling is usually quoted in bit/s. The estimated capacity for the ear is about 8000 bit/s, whilst that of the eye is 3.5×10^6 bit/s. Only a small fraction of this flow of information can be processed because it takes time for the stimulus to activate the sensory organ and for impulses to travel to the brain; it takes time for the brain to identify the signal and initiate the appropriate response; and it takes time to energise the muscles to produce a control movement. During this process, the operator also resolves the uncertainty arising from two sources: firstly, he may not know exactly *when* the signal is coming and therefore when to respond; secondly, in *choice reaction tasks* where different responses have to be made to each of several signals, he may not know *which* signal is coming and therefore which response to make. For a number of signals, including 'no signal', the mean choice reaction time is:

$$t_r = K_r \times \log_2 N_a \qquad \qquad ...[2.3]$$

Mean choice reaction time, s = Constant $\times$ $\log_2$ (Number of alternatives)

For an ordinal decision (yes/no), $\log_2 2 = 1$, so that the constant K_r is the simple reaction time (Fig. 2.15).

This theoretical approach was applied to a study of the mental workload involved in combine harvesting (Fig. 2.16). The number of adjustments were recorded for forward speed, cutter bar height and steering. Both forward speed and cutter bar height were controlled hydraulically and remained at a constant setting unless adjusted by their control lever in one direction to raise the setting and in the other direction to lower it. Steering adjustments were based on the number of steering reversals involving movement in excess of 5° of arc. This gives the choice of 7 control

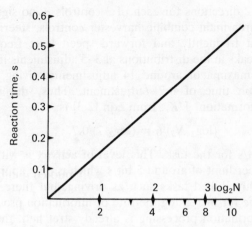

Fig. 2.15 Choice reaction time for a number of alternatives (after Welford, 1968).

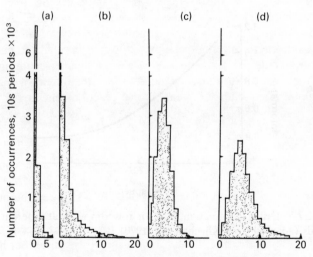

Fig. 2.16 Frequency histograms of combine harvester control adjustments on one machine working for ten days: (a) forward speed control level; (b) cutter bar height control lever; (c) steering wheel reversals; (d) total (after McGechan, 1984).

adjustments (2 directions for each of 3 controls + no signal).

Of these three main combine harvester controls, steering was adjusted most frequently and forward speed least frequently. There were peaks in the distributions at 3–5 adjustments in a 10 s period, with maxima of around 14 adjustments, giving a least choice reaction time of 0.7 s/adjustment. Thus, the rate of processing information, $1/K_r$, from eqn [2.3] is:

$$(\log_2 N_a)/t_t = (\log_2 7)/0.7,$$

or about 4 bit/s for the task. This level of activity is within the suggested upper limit of around 5 bit/s which is only approached by a few highly skilled tasks such as driving, but there is only marginal scope for increasing the rate of information processing. Although information processing is already stretched, the trend towards higher speeds of machine operation rapidly accelerates the rate of receiving information. The net result is a reduction in the average amount of information used by operators at increasing forward speeds as shown in Fig. 2.17 for rowcrop work.

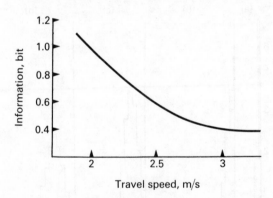

Fig. 2.17 Driver information used for rowcrop work at different travel speeds (adapted from Khachatryan, 1970).

Not only are higher speeds decreasing the time available for processing information, but also the volume of relevant information is increasing because the isolation of the operator in a sound-proofed capsule creates the need for additional displays to monitor both the condition of the vehicle itself and the work

Table 2.8 Relative advantages of real and artificial displays

Real	Artificial
Contain variety of unquantifiable evidence.	Essentially quantitative.

Essentially in real time.	Information can be about:
	  past
	  present
	  future
	Can be:
	  anticipatory
	  time compressed
	  time expanded
	  differentiated
	  integrated

Usually the more reliable.	Subject to failure of sensors and computers.
Observer can fully exercise his selecting and structuring abilities.	Selecting and structuring done by designer.
Maximum versatility.	Often easier to use but less versatile.
Usually less costly.	Can be expensive.
Difficult and expensive to record in detail.	Easy to record.

(*Source*: Singleton, 1974)

output. In agriculture, a large proportion of the control information is obtained visually from the *real display* of the task outputs rather than from an *artificial display* on an instrument panel. The relative merits of real and artificial displays of information is given in Table 2.8. The major opportunity with artificial displays is to reduce the volume of information by electronic data processing prior to display, and even to introduce a greater degree of automation (Fig. 2.18). In combine harvesting, for example, a grain loss monitor can be used to provide performance information for the operator or the information can be linked directly to an automatic travel speed control in which high grain losses reduce travel speed and vice versa.

2.5.2 Human energy expenditure and total stress

As well as the increasing mental workload, agricultural operations also impose varying levels of physical stress. If the oper-

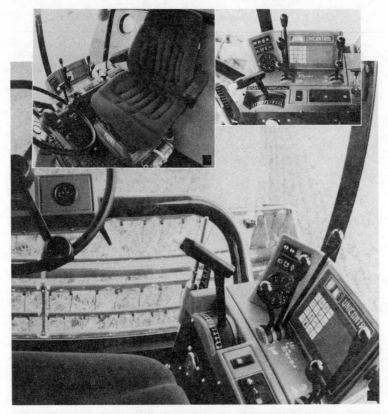

Fig. 2.18 The standard Unicontrol computer monitors combine harvester systems, provides advance warning of malfunction and stops the engine if the warning is overlooked (from Massey Ferguson).

ator's controls are not correctly adapted to his anatomy, the performance expected of him may exceed the limits of tolerance and lead to premature fatigue, impaired health and accidents. Equally, there is a multitude of farm tasks which are not fully mechanised and which may involve heavy physical work. The gross energy expenditure for any activity (including the rest rate) can be measured from the oxygen demand during respiration, whilst the total mental and physical stress can be assessed from the minute-by-minute average of the pulse rate.

In the typical ergonomic data given in Table 2.9 for task

Table 2.9 Human energy expenditure and stress level for various agricultural tasks

Scaling of effort	Task	Gross energy expenditure, W	Mean stress level, heart beat/min
Heavy (540–720 W)	Carrying 40 kg sack of concentrates 15 m and distributing into trough	650	100
Heavy (540–720 W)	Manually forking silage into passage towards animals	560	120
Heavy (540–720 W)	Loading 100 kg sawdust into trolley using shovel	540	144
Moderate (360–540 W)	Pushing trolley of pelleted feed and distributing 3 kg per pig	460	120
Moderate (360–540 W)	Bedding courts with 3 kg straw per animal using bales	420	80
Light (180–360 W)	Loading silage into forage box with foreloader; delivering to trough	190	84
Very light (0–180 W)	Controlling fixed equipment (tower silo unloader and trough conveyor)	120	70

(*Source*: Cermak and Ross, 1980)

durations of 10 minutes, the gross energy expenditure is not related to the mean stress level and foreloader operation is classified as light work. In contrast, continuous foreloader operation in muck handling is considered to be one of the most strenuous activities. This is because all the muscles are under constant stress or 'preload', resulting from the repetitive operation of the many tractor and loader controls which leads to highly fatiguing static muscle strain. In one hour of the muck handling activity, the operator is required to shift gears and operate the clutch 230 times, operate the brakes 100 times and the hydraulic control of the front-mounted loader 250 times. The gross energy requirement can be reduced by careful re-design of the position of the controls in relation to the operator seat and, at the same time, technical innovation may decrease or eliminate the physical stress.

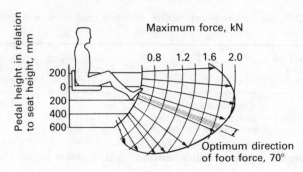

Fig. 2.19 Relationship between the maximum pedal thrust and the height of the seat above the pedal (adapted from Dupuis, 1959).

The maximum pedal thrust is strongly influenced by the difference in height between the pedal and the operator seat, as well as by the distance from the pedal to the seat backrest, without which the thrust is limited to the weight of the leg itself. For foot pedal operation, it is best to have the backrest of the seat at a height equalling that of the pelvis, a maximum of 250 mm high (Fig. 2.19). The distance between the backrest and the pedal should be adjustable to allow the leg to bend slightly at the knee. Between this position and the one in which the leg is stretched out completely (knee straight), there should be an extension of 75–100 mm and this limits the total pedal travel for maximum thrust. In an examination of different vehicles in the 1960s, operating forces on tractor pedals substantially exceeded the physiologically desirable limit of 350 N (Table 2.10). Servo-assisted brakes, as fitted to trucks at that time, required much lower pedal operating forces. Further technical developments

Table 2.10 Brake pedal forces and resulting vehicle deceleration

Vehicle	Speed before braking, km/h	Maximum deceleration, m/s^2	Pedal thrust, N
Tractor 15 kW	20.0	1.66	1 020
Tractor 9 kW	17.7	2.58	970
Truck 100 kW	40.0	3.30	310

(*Source*: Dupuis, 1959)

have made it possible to eliminate the foot brake altogether by adopting a hydrostatic transmission system. This stepless hydraulic unit provides single lever control of vehicle movement from full speed forward to full speed reverse with powerful brakes in between at zero speed. The simplicity of combining the clutch, the brake and the gear lever into a single hand control is offset by the higher cost and lower transmission efficiency, and can only be justified for operations where the shuttle reverse is a major advantage. Nevertheless, it illustrates the alternatives that are available to limit the total level of stress imposed on the operator.

Applying *anthropometric* data, it is possible to produce graphical layouts of the workplace for optimum positioning of all the hand and foot controls (Fig. 2.20).

● **Anthropometry** is the study of the human body, of the dimen-

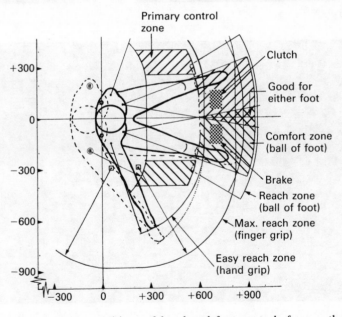

Fig. 2.20 Optimum positions of hand and foot controls from anthropometric data using two shoulder (S) and one hip (H) locations for both a 97.5 percentile male and a 2.5 percentile female (adapted from Purcell, 1980).

(a)

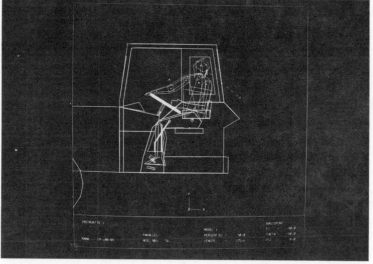

(b)

Fig. 2.21 Sideviews of tractor cab; (a) operator turning backwards; (b) anthropometric model in same posture (from Zegers, 1985).

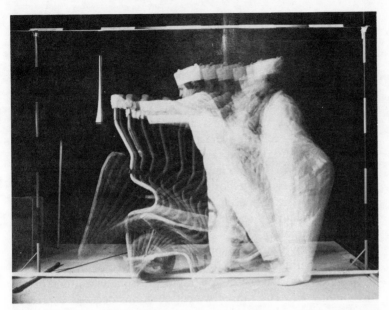

Fig. 2.22 The use of interrupted light photography to provide a set of space envelopes for individual body movements which are combined into a single kinetosphere (from Scottish Farm Buildings Investigation Unit).

sions of large, average and small men and women and the limits of their body movements and strength.

This type of design work is simplified by the use of computer graphics to visualise the shape of the body and to simulate body movements in the workplace (Fig. 2.21). The field of vision of the operator can also be explored and the support pillars for the roof of the tractor cab relocated to minimise blind spots.

On leaving the confines of the operator capsule on a machine, it becomes more difficult to prescribe the limits of the workspace, since the operator may take up a multiplicity of positions in carrying out a task. Using interrupted light photography, a set of space envelopes for individual body movements can be combined to provide a composite kinetosphere (Fig. 2.22). Superposition of the individual kinetospheres for all the envisaged tasks – for example, carrying bales, distributing feed and cleaning – establishes the operator's spatial needs and assists in the rational design of feed passages in buildings.

2.6 OPERATOR EXPOSURE TO NOISE AND VIBRATION

2.6.1 Noise

Apart from being an annoyance, noise may interfere with operator efficiency and, by hindering communications, be the cause of accidents. Most important of all, it can cause damage to hearing. The risk of deafness depends on the quantity of the sound energy received over a period of time. A temporary hearing loss, lasting from a few seconds to a few days, may result from exposure to intense noise for a short time. Much more serious is the regular exposure to high noise levels over long periods such as experienced by tractor drivers. By damaging the inner ear structure through excessive noise, the loss of hearing becomes incurable.

- **Noise** is any *unwanted* sound.

Sound is generated by the vibration of surfaces, or by turbulence in an airstream, which sets up rapid pressure vibrations in the surrounding air. The rate at which vibrations occur is expressed as a frequency. The simplest vibration is a pure tone which may be represented by a sinusoidal curve (Fig. 2.23) with the frequency:

$$f = 1/T_p \qquad \qquad \ldots[2.4]$$

Frequency, Hz = 1/Period, s

The amplitude of the vibration is usually expressed as the *root mean square* (r.m.s.) value because of its direct relation to the energy content of the signal in a linear system; thus, for a simple pure tone:

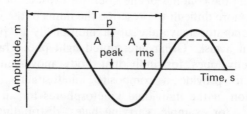

Fig. 2.23 A sinusoidal vibration with the peak and the root mean square amplitudes and the period of the wave.

$$A_{rms} = 1/(A_{peak} \times \sqrt{2}) \qquad \ldots[2.6]$$

Root mean square amplitude, m $= 1/($Peak amplitude, m $\times \sqrt{2})$

As airborne sound is a variation in the normal atmospheric pressure, the root mean square of the airborne sound pressure, relative to a reference sound pressure of 0.000 02 N/m^2, can be expressed as a *sound pressure level* on a logarithmic scale in decibels.

Mathematically, the **sound pressure level**, *SPL*, is:

$$SPL = 20 \log_{10}(P_{rms}/P_{ref}) \qquad \ldots[2.7]$$

Sound pressure level, dB $= 20 \log_{10} \dfrac{\text{Measured r.m.s. sound pressure, N/m}^2}{\text{Reference sound pressure, N/m}^2}$

- The **audible frequencies** are the range of frequencies detected by the human ear.

These extend from 20 Hz to 20 000 Hz, but there is greater sensitivity to sounds with frequencies between 1000 Hz and 4000 Hz. When monitoring noise levels in an operator work-space, it is usual to use a meter which incorporates an '*A-weighting filter*'. This filter accepts all audible frequencies but makes the meter less sensitive to the very low and very high frequencies which have less effect on the operator.

Using a logarithmic scale for sound measurements means that every increase of 10 dB(A) makes the sound seem about twice as loud but causes a tenfold increase in the rate at which the sound energy is recorded. Typical values for sound pressure levels as heard by a bystander are:

rustle of a leaf   –   10 dB(A)
normal conversation –   60 dB(A)
busy street noises   –   70 dB(A)
chain saw   – 110 dB(A)

A 3 dB increase in noise gives only a small apparent change in loudness of a sound but doubles the rate at which the sound energy is received. Thus, the time an unprotected person can safely stay in the noise is halved (Table 2.11).

Approved 'Q' safety cabs now fitted to tractors ensure that the noise level at the operator's ears does not exceed the statutory limit of 90 dB(A) and, in many of them, the noise level has been halved down to 87 dB(A). Working with the cab window open,

Table 2.11 Levels of noise which indicate a serious hazard to hearing

Exposure duration, hours per day	Maximum sound level, dB(A)
8	90
4	93
2	96
1	99
$\frac{1}{2}$	102
$\frac{1}{4}$	105

however, can increase the noise level substantially above the limit. Buildings can also intensify noise from fixed equipment and even animals can create noise levels above 110 dB(A) especially at feeding time. In these circumstances, personal ear protection should be worn. Ear muffs are the most effective form of protection for they effectively reduce harmful noise levels but still permit the wearer to hear verbal messages and machine warning signals.

2.6.2   Vibration

Exposure of the human body to vibrations not only reduces operator comfort but also can significantly impair the ability to see detail quickly and to make precise control adjustments. Many tractors are operated at less than two-thirds maximum power because of the operator's unwillingness to withstand vibrations at full speed. Continuous exposure to the vibration from early designs of chain saws caused permanent loss of feeling in the fingers, the 'white finger syndrome'.

● **Vibration** is oscillatory motion of a mechanical system.

Vibration may be split up into three frequency bands:

0–15 Hz	for ride vibration;
15–50 Hz	for mechanical vibration from engine and power transmission, corresponding to rotational speeds from 900 to 3000 rev/min;
50 Hz–20 kHz	for noise.

The oscillatory motion can be extremely complex but considering only simple harmonics, the displacement velocity and acceleration functions of the wave form, as shown in Fig. 2.23, can be mathematically described as:

$$X = A_{peak} \times \sin(\omega \times t) \qquad \ldots [2.8]$$

Displacement, m   = Peak amplitude, m
                       × sin(Frequency of vibration, rad/s × Time, s)

$$v = \omega \times A_{peak} \times \cos(\omega \times t) \qquad \ldots [2.9]$$

Velocity, m/s   = Frequency of vibration, rad/s × Peak amplitude, m
                       × cos(Frequency of vibration, rad/s × Time, s)

$$a = -\omega^2 \times A_{peak} \times \sin(\omega \times t) \qquad \ldots [2.10]$$

Acceleration, m/s^2 = − (Frequency of vibration, rad/s)^2 × Peak amplitude, m
                       × sin(Frequency of vibration, rad/s × Time, s)

The level of vibration is usually measured as the root mean square of the acceleration such that:

$$a_{rms} = \omega^2 \times A_{peak}/\sqrt{2} \qquad \ldots [2.11]$$

Root mean square acceleration, m/s^2 = (Frequency of vibration, rad/s)^2
                             × Peak amplitude, m/$\sqrt{2}$

It is also convenient to express acceleration in terms of the acceleration due to gravity by dividing the root mean square acceleration by 9.8 m/s^2.

The human tolerance of vertical vibration is shown in Fig. 2.24 for various durations, but in all cases tolerance in the 4 to 8 Hz range is always least. These acceleration values for the *fatigue-decreased proficiency boundaries* should be divided by 3.15 to obtain the *reduced comfort boundaries*; whilst the *upper exposure limit* is twice the fatigue-decreased proficiency boundary.

The typical vibration levels in practical agriculture which are given in Table 2.12 are high in relation to the International Standard. These ride-induced vibrations do appear to have ill effects on the tractor driver as indicated in a study of the incidence of spine deformations of tractor drivers compared with the population as a whole (Fig. 2.25).

The main design emphasis to reduce vibration has been in the development of sprung seats but this can only decrease the vertical component, whereas pitch and roll are also important. Even with moderately good suspension seats, the fatigue-decreased proficiency limit can be reached in as little as one hour

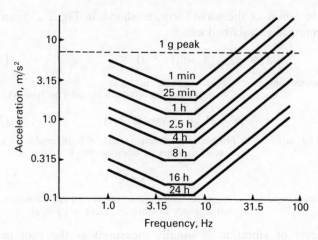

Fig. 2.24 Fatigue decreased proficiency boundaries for vertical acceleration (after ISO 2631).

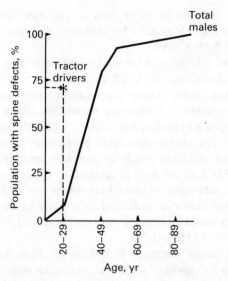

Fig. 2.25 The incidence of spine deformations for tractor drivers compared with the male population as a whole (after Rosseger and Rosseger, 1960).

Table 2.12 Average tractor driver ride vibration levels in order of magnitude (vertical vibration) with the exposure limits from ISO 2631 shown for comparison

Operation/machine	Average vibration level (frequency weighted), m/s^2		
	Vertical	Lateral	Longitudinal
Rotary cultivator	0.152	0.132	0.190
Baler	0.513	0.569	0.702
Roller	0.533	0.372	0.387
Loader	0.613	0.459	0.510
Potato spinner/harvester	0.617	0.643	0.183
Manure spreader	0.690	0.507	0.695
Disc harrow	0.706	0.730	0.631
Mouldboard plough	0.716	0.524	0.504
Light cultivator	0.749	0.536	0.571
Heavy cultivator	0.772	0.503	0.540
Fertiliser spreader	0.835	0.617	0.727
Trailer	0.838	0.571	0.785
Mower	0.851	0.407	0.424
Drag harrow	0.865	0.365	0.250
Sprayer	0.871	0.546	0.544
Drill	0.959	0.533	0.678
Hay turner/tedder	1.319	0.630	0.533
Maximum exposure level 8 h/day	0.70	0.45	0.45
Maximum exposure level 2.5 h/day	1.60	1.00	1.00

(*Source*: Stayner and Bean, 1975)

at elevated speeds. If operators are to be subjected regularly to such conditions for the working day, a much lower suspension natural frequency than that attainable with sprung seats must be employed. Thus, research is well advanced with the design of a complete suspension system for the operator capsule, giving isolation for the operator from noise and vibration at one and the same time.

2.7 WORKSPACE ENVIRONMENT

For maximum operator productivity, the immediate workspace must conform to the expected standards of comfort in terms of the climatic environment and must minimise atmospheric pollution from dust, smells and noxious chemicals.

2.7.1 Climatic environment

The physiology of comfort involves a complex interaction of a number of factors, including:
○ **mean radiant temperature;**
○ **ambient air temperature;**
○ **relative humidity;**
○ **relative air velocity;**
○ **thermal resistance of clothing;**
○ **metabolic rate of operator.**
On a bright sunny day, the mean radiant temperature in a tractor cab may be 5 to 10 °C above the cab air temperature. Forced ventilation, with an air velocity of between 1 and 3 m/s, plays an important role in reducing the effect of solar radiation. In order to maintain the thermal equilibrium in winter, however, the workspace temperature must be adjusted to balance the varying metabolic rate of the operator which may range from 140–350 W and the varying thermal resistance of his clothing which may have values of 1.0, 1.5 and 2.0 clo for light, medium and heavy materials (1 clo $\simeq$ 0.16 °C m^2 W^{-1}). The lighter his clothing and the lower the ambient air temperature, the higher the optimum workspace temperature required (Fig. 2.26).

Operator comfort is more correctly specified by a combination of dry-bulb temperature and relative humidity to give an *effective temperature*.

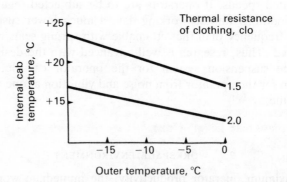

Fig. 2.26 The effect of outside temperature and thermal resistance of clothing on the optimum temperature within a tractor cab (after Eriksson, 1973).

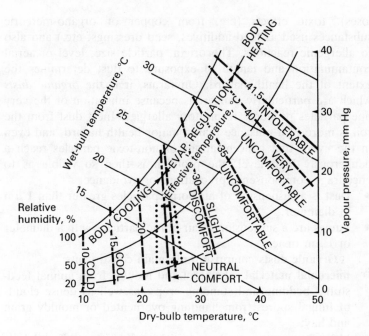

Fig. 2.27 The thermal comfort zone, shown dotted for a sedentary person, and the discomfort limits based on the effective temperature scale which empirically combines the interaction between dry-bulb temperature and relative humidity (adapted from ASHRAE, 1985).

- The **effective temperature** is an empirical index representing combinations of dry bulb temperature and relative humidity that produce the same level of *'skin wettedness'* as caused by regulatory sweating.

In the ASHRAE Comfort Chart (Fig. 2.27), the effective temperatures for the indoor thermal environment are related to the psychrometric parameters. Caution must be used in transposing these data from buildings to agricultural machines because it is assumed that the radiant temperature is nearly equal to the dry-bulb temperature and the air velocity is less than 0.25 m/s.

2.7.2 Airborne dusts

Inhalation of dust can lead to respiratory disorders (pneumoconioses), toxic effects (e.g. from copper or organo-mercuric substances used as feed additives, seed dressings, etc.) and also to allergenic reactions. The origin, particle size, level of aerial contamination and length of exposure to dust determines the extent of the health hazard. On farms, it is the *organic dusts* which are particularly dangerous because inhalation of the very fine particles produces respiratory allergies. Inert dust from the soil is more of a nuisance than a major health hazard, and even in the event that concentrations of non-toxic particles reach a dangerously high level, the problem is then so visible as to encourage commonsense precautions or avoidance.

- **Dust** is a suspension of air-borne particles greater than 1 μm in diameter.
- **Fumes** are a suspension of air-borne particles with a diameter of 1 μm or less.

  Organic dusts comprise three main constituents:
- **microbial material** – bacteria and moulds from animal feed-stuffs, bedding and standing crops, as well as dense clouds of fungal spores from handling overheated or mouldy grain and hay;
- **plant fragments** – pollen, particles of grain, chaff and hay, and dried faecal matter;
- **animal fragments** – particles of skin, feather, hair, and mites which can reach high populations on mouldy feeds.

Pollen particle size ranges from 4 to 80 μm and fungal spores may be as small as 5 μm. As there is no standard dust classification system, it is convenient to divide dust into the *respirable fraction* and the *non-respirable* fraction.

The **respirable fraction** consists of particles smaller than 5–8 μm and these are so fine that they can penetrate into the deepest recesses of the lung and lodge in the alveoli. Spores in contact with the lung tissue can cause sensitisation and the formation of antibodies. The allergic response varies with the immunological reactivity of the subject and the degree of exposure. The onset of symptoms (breathlessness, fever and eventually weight loss) can often be delayed several hours. Acute alveolitis leads to fibrosis of the lung – the condition known as farmer's lung – through exposure to large numbers of pathogenic spores (1.4 $\times$ 10^6 spores/m^3 of air) of *Aspergillus fumigatus*.

The **non-respirable fraction** consists of dust particles greater than 5 μm diameter and these are usually deposited along the upper part of the respiratory tract, i.e. the nose, throat, trachea and bronchi. These large particles are trapped by the minute hairs or cilia, constantly in motion, which move the dust particles towards the nose and mouth or adhere to mucous-covered walls of the bronchi. This self-cleaning action ensures that most of the larger particles do not reach the lungs unless the level of concentration exceeds the 'filter capacity'.

A dust concentration of 10 mg/m^3 of air has been suggested as a general nuisance threshold but, in agriculture, this limit is often exceeded when tending livestock (Table 2.13). In combine

Table 2.13 Dust concentrations in livestock production operations

Task	Location	Dust concentration, mg/m^3	Air velocity, m/s
Cleaning and bedding operations			
Sweeping silage and concentrates in feed pass	Diary cattle house	1.9	0.3
Sweeping floor	Calf house	9.0	0.3
Dusting fittings and walls, floor swept and vacuum cleaned	Poultry house	76	0.2
Bedding with sawdust	Cow cubicle house	4.5	0.3
Bedding with baled straw	Cattle court	12	0.4
Muck removal under cages	Poultry house	15	0.2
Feeding operations			
Dispensing silage, draff and concentrates by mixer wagon	Cow cubicle house	3.7	0.3
Forking hay into heck	Calf house	13	—
Feeding concentrates by bucket	Slatted cattle court, Pig Weaner house	21	0.1
Feeding meal by bucket	Pig weaner house	49	—
Wet feeding	Pig fattening house	8.6	—

Table 2.13 (cont'd)

Task	Location	Dust concentration, mg/m^3	Air velocity, m/s
Feed processing operations			
Filling concentrates bin	Feed store	280	—
Rolling barley: without dust extractor	Feed store	100	—
with dust extractor		9.0	0.1
Produce collection operations			
Milking	8-stall rotary-tandem parlour	1.6	—
Egg collection	Laying house with 3-tier cages	25	0.1
Background dust levels without human activity			
—	Cow cubicle house	1.1	0.1
—	Milking parlour	0.8	—
—	Cattle court	2.2	0.3
—	Meal store	3.3	—
—	Poultry laying house	3.7	0.1

(*Source*: Cermak and Ross, 1978)

harvesting, average dust levels range from 2 to 6 mg/m^3 but can exceed 15 mg/m^3 under certain conditions.

Operator protection from dust can be achieved in three ways:
○ **dust prevention** – repair dust leaks, e.g. in feed processing equipment, and minimise dust generation by using augers in place of pneumatic conveyors;
○ **dust capture** – locate duct inlets close to dust sources since the minimum capture velocity of 0.25 m/s for a circular duct with a transport velocity of 5 m/s occurs at a locus which is the surface of a sphere of radius of only $2\frac{1}{2}$ times that of the duct;
○ **personal protection** – provide workspace with air filtration and pressurisation of 50–100 Pa above atmospheric to

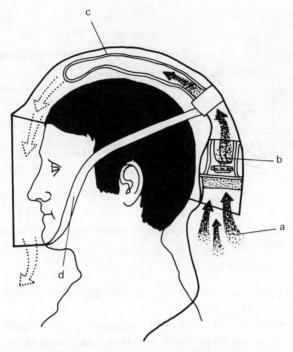

Fig. 2.28 Positive pressure powered respirator helmet in section, showing: (a) air intake; (b) electric motor; (c) filter; (d) sealed edge, with arrows indicating the course of airflow (from Watson, 1979).

prevent dust entry or wear a suitable dust respirator (Fig. 2.28).

2.7.3 Gas hazards

Vehicle exhaust gases can be dangerous in confined spaces where fork lift trucks are handling pallets in and around vegetable stores and where two-wheeled tractors are being used in glasshouses. With diesel engines, the most serious pollutants are likely to be oxides of nitrogen. Nitrogen dioxide has an insidious poisoning effect at concentrations which are so low that precise measurement is difficult. With petrol engines, concentrations of carbon monoxide during normal cultivation work under glass varied from

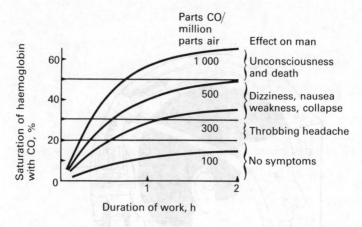

Fig. 2.29 Effect of human exposure to carbon monoxide when working (adapted from Spencer, 1961).

40 to 400 parts per million (p.p.m). At these levels, the health risk is serious (Fig. 2.29).

Concentrations of hydrogen sulphide gas released from agitated slurry can be in excess of 2000 p.p.m., more than enough to cause immediate collapse, even after a single breath. Other than at low concentrations, characterised by the strong 'rotten eggs' smell, hydrogen sulphide stuns the olfactory nerves, giving victims no warning of the danger. Flammable methane gas is also produced by slurry and can accumulate, so that the use of naked lights should also be avoided.

In high moisture grain silos and haylage towers, carbon dioxide accumulates and depletes the available oxygen. Carbon dioxide is much denser than air and reaches asphyxiating concentrations at the base of the tower. An operator descending into the layer is rapidly suffocated.

With all these gas hazards, the main precaution is to provide good natural ventilation of the area or forced ventilation in confined spaces prior to entry. *No* respirator will give protection; only breathing apparatus is adequate where the gas pocket has not been dispersed.

2.7.4 Odour nuisance

With the intensification of the livestock industry, the problem of unpleasant odours has become more acute. Whilst land spreading of slurry is the most frequent source of complaint, odour nuisance can be caused through silage storage and through rendering animal products. The extent of the nuisance of the odour depends on its offensiveness, its strength and duration, the proximity of populated areas, and on the wind direction and strength.

Odour level can only be assessed using the sensory methods based on the human nose. Odours collected by a standard bag sampling technique are presented in a range of dilutions to members of a test panel. By recording the panel response, the odour threshold is determined on the number of dilutions of the sample with clean air which will result in a 50 per cent level of detection by the odour panel.

Despite the difficulties of odour level assessment, the primary substances causing odour nuisance are not in dispute; they are:

○ **hydrogen sulphide;**
○ **ammonia;**
○ **aliphatic aldehydes;**
○ **mercaptans;**
○ **amines.**

As fine dust particles are primarily responsible for transporting obnoxious odours, air scrubbers, where the contaminated air is blown over a water-sprayed honeycomb, are relatively efficient devices for odour control in livestock houses. Biological air scrubbers, consisting of a peat matrix on which a bacterial floc is established, have been very successful in ameliorating the offensive odours from reducing animal carcases to tallow and bone meal, a process that most people prefer to avoid!

The annoyance caused during land application of wastes can be minimised by choosing weather conditions favourable to rapid dispersion of odours (Table 2.14). Odours disperse most slowly on clear nights with light winds, when the atmosphere is in a 'stable' condition; slow moving air will be mixed with only a shallow depth of air above. Unstable conditions, in which odours are diluted rapidly, occur on brighter days, especially if associated with strong winds; air mixing will be much greater and may be over a layer of about 1500 m high.

Table 2.14 Conditions favourable for dispersal of odour, the highest numerical rating indicating fastest dispersion

Wind speed	Night dispersion of odour			Day dispersion of odour		
	Clear sky	Broken cloud	Overcast	No sun	Weak sun	Bright sun
Light	1	1	2	2	3	4
Moderate	2	2	3	3	4	5
Strong	3	3	4	4	5	5

(*Source*: Bird, 1978)

The frequency of spreading is also important, with the majority of people – operators and public alike – accepting a period of strong smells for a few days due to intense spreading activity, in preference to intermittent odours associated with weekly handling. In particularly sensitive areas, slurry injection into the soil is an effective method of odour control.

2.8 PREVENTION OF ACCIDENTS

Health and safety risks are greater in agriculture than in some other industries. With respect to employees, the risk of a fatal accident, using 1979 figures, is 17 per 100 000 persons employed (compared with 19.8 per 100 000 persons in the mining and petroleum industries).

A distressing component of the farm accident record is the fatal and serious injuries sustained by children, reflecting the proximity of home and workplace on the 'family farm'. There are also health hazards due to the particular nature of farm work: the skin disorders and burns through widespread use of agrochemicals aerial spraying and straw burning; and the risk of tetanus through wounds infected by the bacterium *Clostridium tetani* from the soil, farmer's lung from organic dusts (see section 2.7.2) and Weil's disease caused by a spirochaete excreted in rats' urine.

In addition to the compelling humanitarian reasons for making agriculture a safer and healthier place in which to live and work, there are economic reasons for pursuing the same goal. Accidents cost money and resources in the form of loss of income, loss of

Table 2.15 Cost elements of non-fatal reported accidents by accident type

Accident type	Number of accidents	Percentage of resource cost due to				Average resource cost £
		Medical cost	Damage cost	Delay cost	Delay avoidance cost	
1. Tractor overturn	33	7	60	17	15	3120
2. Tractor p.t.o.	6	16	0	80	4	340
3. Other tractor	67	21	24	46	10	1110
4. Other self-propelled machine	43	13	17	49	22	1550
5. Field machine p.t.o.	7	13	0	65	22	670
6. Other field machine	181	19	1	65	15	890
7. Hand tools	66	25	0	52	23	610
8. Stationary machinery	61	21	0	59	19	860
9. Circular saws	29	26	0	64	11	840
10. Electrical	4	18	0	50	33	330
11. Falls	105	18	0	50	32	760
12. Bulls/boars/other	22	30	0	39	31	1080
13. Other cattle	58	12	0	52	35	370
14. Poisoning	5	14	0	53	33	560
15. Falling/swinging objects	39	27	12	39	23	650
16. Strains/wounds	37	12	1	60	28	300
17. Weil's disease	1	60	0	40	0	2370
18. Farmer's lung	5	40	0	43	17	1950
19. Other	22	34	2	45	19	610
Overall	791	18	13	49	19	910

(*Source*: Monk *et al.*, 1984)

production, damage to property, and payment of injury benefits and medical expenses to the injured party (Table 2.15). Equally, of course, the commitment of resources for accident prevention must remain cost-effective and, for this reason, greater attention is given to reducing accidents during machinery operations.

Tractors are basically stable machines but they are sometimes used in circumstances where there are risks from overturning. With the provision of roll-over protection, the operator is now unlikely to be crushed in an overturning accident provided he stays inside the safety cab. It is not as easy, however, to ensure that he does not sustain incidental injuries from being thrown about inside the cab. Thus, the main emphasis is on accident prevention through safe tractor operation. In a survey of tractor overturning accidents, the main causes were attributed, firstly, to exceeding the tractor operating limits and either tipping over (stability loss) or sliding downhill (control loss) and, secondly, to driver's misjudgement (Table 2.16). Instrumentation is available to monitor slope angle and provide audible warning of dangerous conditions but is not yet widely used, so excessive reliance is placed on driver experience. A graphic example of the lack of time available to prevent an accident is shown in Fig. 2.30, whilst safe tractor operation is described more fully in section 8.

Table 2.16 Classification of tractor overturning accidents according to accident cause

Category	Cause	Number	%
1. Tractor related (i.e. tractor limitations exceeded)	Stability loss		
	(i) Slope exceeds tip angle	95*	17
	(ii) Speed high	56*	10
	(iii) Ground rough	34*	6
	Control loss	125*	22
2. Driver related	Driver's misjudgement	145	26
3. Miscellaneous	Varied causes but including traffic accidents (28) and driverless tractors (37)	105	19
Total		560	100

* Tractor limitations exceeded in 55% of total accidents.
(*Source*: Owen and Hunter, 1983)

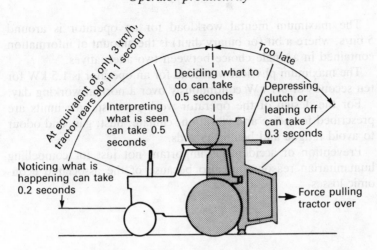

At equivalent of only 3 km/h,
tractor rears 90° in 1 second

Too late

Deciding what to
do can take
0.5 seconds

Depressing
clutch or
leaping off
can take
0.3 seconds

Interpreting
what is seen
can take 0.5
seconds

Noticing what is
happening can take
0.2 seconds

Force pulling
tractor over

Fig. 2.30 Accidents happen and an experienced driver who may need 1.5 seconds to decide on and carry out remedial action has insufficient time to respond to a tractor rearing over backwards (after HSE, 1980).

2.9 SUMMARY

For farm planning purposes, annual labour requirements are calculated in standard man-days for individual enterprises.

Work study, comprising both method study and work measurement, is used to examine the operational efficiency of an activity in order to achieve improvement.

The systematic appraisal of an activity by method study involves six stages – select, record, examine, develop, install and maintain – with relevant data about the activities being recorded on flow process charts, multiple activity charts, and flow and string diagrams. Critical examination of the data is conducted by means of a questioning sequence.

The principles of motion economy form the basis on which all movement patterns should be constructed.

Network analysis is a planning and scheduling technique to ensure that activities are completed in the correct sequence and with the minimum resource use.

Ergonomics is the study of the man/machine interaction to improve the design of the workplace environment.

The maximum mental workload for an operator is around 5 bit/s, where a bit (or binary digit) is the amount of information contained in a single choice between two alternatives.

The maximum physical workload for an operator is 1.5 kW for ten seconds or 0.1 kW continuously over a normal working day.

For the design of the operator workspace, specific limits are prescribed for noise and vibration, climate, dust, gas, and odour to avoid fatigue and health hazards.

Prevention of accidents is important not just for compelling humanitarian reasons but also because of the associated economic losses.

3

MACHINE PERFORMANCE

OUTLINE

Machine capacity; Spot, overall, seasonal and calendar rates of work; Field efficiency; Effective width of machine; Filling and unloading procedures; Turning techniques; Fieldwork patterns, circuitous and headland; Headland width; Optimum land size; Field size and shape; Machine blockages; Breakdowns and repairs.

APOSTROPHE – TWAL OWSEN PLOUGH

'The plough itself is beyond description bad! Yet, rude as the plough was and imperfect its equipments, under the steady persevering pull of a team of great sinewy oxen of six or seven years old, so long as it could be kept in the ground it made a large if unshapely furrow, turning over or pushing aside a mass of soft earth and clearing away earth fast stones that stood in its way, aye until the draught chain broke or other part of the gearing gave way.'

As the tenants in common provided the livestock, there was no lack of animal power and up to twelve oxen were forced into a huge cavalcade without reference to the nature of the ground. The foremost pair which were called the *on wyner* and *wyner* literally held a pivotal position 'in so far as turning the unwieldy team on a moderate width of end-rig depended on their easing off the draught gradually and featly'. And then, as the young oxen were always trained immediately behind the leaders, in the *on steer* draught and *steer* draught position, 'it depended on the trustworthiness of the veteran wyners that these juniors should be kept steady, and prevented running into untimeous and uncanny escapades, even to the extent of breaking their

harness at times, and scampering off from the draught altogether'. Following the juniors were the *fore throck*, *mid throck*
and *hind throck* pairs, and, finally, the sixth pair were the *fit
on land* and the *fit in fur*. A yoke which lay across the necks
of each pair of oxen was attached to a bow of ash surrounding
every separate ox's neck. And the *fit on land* ox was not
considered fully trained until it had learnt to lower its neck
when the ploughman wanted the plough to go a little deeper!

'The numbers of animals employed in ploughing and the
rough condition of the ground called for the presence of a
driver who looked in advance for half concealed stones and
backwards to extricate the ploughman from difficulties. At the
nose of the plough walked a second man who carried a pole
hooked to the point of the beam which enabled him to push
the plough towards the furrow if the slice was too broad or
draw it towards the land if too narrow. He also helped to
regulate the depth of the furrow. A fourth man often followed
the plough with a spade to compress the bold shoulder of earth
thrown up by the heavy plough, break the clods and turn over
slices that had fallen back and such parts of the ground as the
plough had missed. This formidable procession of animals and
men moved forward at funereal pace to the accompaniment of
much shouting and whistling – "a hideous Irish noise" – and
scratched less than half an acre a day. Stoppages were frequent
on account of the rough going. The earth-bound stones that lay
hid below the surface and firmly fixed into it were the very bane
of tillage in older times and occasioned the destruction of nine
out of ten of all ploughs that were made however strong and
heavy they might have been'.

'A field ploughed with this machine looks as if a drove of
swine had been moiling it'. The run rigs were ploughed length-
ways into ridges through the practice of always gathering soil
towards the crown which was three or more feet high and from
ten to twenty six feet broad at each end, but bulging out to
thirty or forty feet at the middle. They were crooked like a
prolonged S, not for a set purpose, but simply due to the
difficulty of keeping the heavy cumbersome plough and its
lumbering cavalcade of oxen on a straight course over a length
of 200 to 500 yards. The serpentine path of the ridges not only
made the labour of ploughing and laying the earth over more
difficult but also caught and retained surface water in its
winding furrows. Moreover, this method of gathering the soil
tended both to make the crown too dry with the best of the

crop on the exposed heights liable to be laid by the wind and to cause the lower half of the ridge with little friable soil, soured by poor drainage, and shaded from their full share of the sun. 'Taking one year with another the quantity of weed seeds must be nearly equal to that of the grain produced.'

(Sources: Handley; Anon 1877)

3.1 INTRODUCTION

Efficient machinery management requires accurate performance data on the capability of individual machines in order to meet projected work schedules and to form balanced mechanisation systems by matching the performance of separate items of equipment. Whilst engineering specifications are always precise and readily available, field performance data are sparse for a variety of reasons. There is considerable variation in operating conditions, such as in topography, surface roughness, stoniness, and soil trafficability; there is variation in the processed material, be it soil strength affected by mechanical analysis or crop density varying with yield and both further influenced by moisture content; and there is variation in the performance measures which are used.

Machines can be evaluated over a short period in productive work – equivalent to speed trials with a flying start – or they can be monitored over time, taking into account associated delays – equivalent to the Le Mans 24 hour Race. The former identifies peak performance, whilst the latter places greater emphasis on driving skill, machine reliability and team organisation. Attaining consistently high machine performance over an extended period in the field demands driving skill to eliminate time-wasting manoeuvres, good team organisation in materials handling procedures which are a major component of field work and rationalisation of field size, shape and work patterns.

3.2 MACHINE CAPACITY

The performance of an agricultural machine is assessed by the *rate* at which an operation is accomplished and by the *quality* of the output. The significance of these performance measures in farming differs from that for manufacturing industries. The rate of work is important because of the seasonality of field operations and because of the vagaries of the weather. The perishable and fragile nature of farm produce also emphasises the need for careful and gentle machine operation to minimise crop damage. These factors determine the upper limit for the rate of machine performance, the design (or rated) *capacity* of the machine.

Machine capacity is specified as a quantity per unit time, and includes:

○ **area capacity;**
○ **commodity throughput capacity;**
○ **total throughput capacity.**

3.2.1 Area capacity

Area capacity is used to identify the rate of work for field operations where the area of ground covered is the simplest measure of work achieved, such as in tillage and spraying.

● The **theoretical area capacity** is the maximum possible area capacity at a given operating speed and fully utilising the operational width of the machine.

In practice, however, machine bouts are overlapped to give an effective width which is 2–5 per cent less than the maximum, so the *effective area capacity* is also less. Consider, for example, the essential information for a spot check of a combine harvester, equipped with a 6 m width of header:

Test duration   = 1 min
Distance travelled   = 100 m
Effective width of cut   = 5.85 m
Theoretical area capacity   = speed × header width
$$= \frac{100 \times 6 \times 60}{10\,000} = \underline{3.6 \text{ ha/h}}$$
Effective area capacity   = speed × width of cut
$$= \frac{100 \times 5.85 \times 60}{10\,000} = \underline{3.51 \text{ ha/h}}$$

3.2.2 Throughput capacity

For harvesting operations, *commodity throughput capacity*, in tonnes per hour of produce (or number of bales per hour for a baler, and so on), is a better overall indicator of machine performance because it accommodates the variability in crop yield. Commodity throughput capacity, however, only accounts for machine performance in handling the saleable portion of the the crop. *Total throughput capacity*, again in tonnes per hour, is used to indicate the performance in terms of total material flow through a machine, such as a combine harvester or potato harvester which separate a saleable product from crop residues or soil contamination.

In combine harvesting, for example, the rate of work is influenced not only by grain yield but also by the yield of 'material other than grain' (straw, chaff, etc.). If, during the spot check in section 3.2.1, some further information was collected, namely:

weight of grain harvested = 250 kg
grain : straw ratio = 1 : 1

then the throughput capacities can be determined.

Commodity throughput capacity = grain harvested per hour
$$= \frac{250 \times 60}{1000} = \underline{15 \text{ t/h}}$$

Total throughput capacity = grain and straw harvested per hour
$$= \frac{(250 + 250) \times 60}{1000} = \underline{30 \text{ t/h}}$$

With a crop such as flax, for instance, it is the 'straw' which is the saleable produce instead of the seed.

Machine performance for potato havesting is also governed by the rate of handling a total number of objects, irrespective of whether they are stones, clods or potatoes. For a potato harvester with a constant total throughput capacity, the commodity capacity in stone-free conditions may be as much as three times that in a very stony soil.

3.2.3 Rates of work

The various capacities just calculated are *spot rates of work* because the test, or spot check, was of very short duration and was conducted when the machine was engaged in productive work.

- **Spot rate of work** is the performance of a machine in productive work.

The spot rate of work accounts for the most important time element involved in machine operation, the time when the machine is 'earning its keep'. It is, therefore, the obvious choice for performance comparisons between machines because it excludes any distortion from organisational delays or from side effects, such as turning time for different field sizes. Just as the pit stop can make or break a grand prix racing champion, so field operations involve routine interruptions to the productive work.

The *overall rate of work* is less than the spot rate of work because it takes account of these routine interruptions for turning or stopping at the end of each bout across the field. This performance rate includes a further three time elements, in addition to the time element in productive work used to calculate the 'spot' rate (Table 3.1).

- The **overall rate of work** is the performance of a machine over the complete operational cycle (productive work time plus routine interruptions for turning and product handling).

Farmers instinctively think in terms of the 'overall' rate. With any luck, it is the amount of crop harvested from the time a combine harvester goes into the field until it comes out again. It is entirely realistic to maintain the overall rate of work over long periods when things are running smoothly, and this makes it the figure to use when planning the grain haulage and grain drier intake arrangements associated with the combine harvesting operation.

The *seasonal rate of work* is a more specialised notion. It includes major breakdowns, mealtimes, travelling from field to field and any other delays which prevent the machine from working while weather and soil or crop conditions would otherwise be suitable (Table 3.1).

- The **seasonal rate of work** is the performance of a machine over the available time (the total duration of the operation excluding only weather delays).

Seasonal rates of work are relevant for forward planning of work schedules.

For a combine harvester with a 3 m wide header and an average operating speed of 5 km/h in the crop, typical values are given for the effective spot rate of work, the overall rate of work and the seasonal rate of work in comparison with the theoretical

Table 3.1 Time elements associated with different rates of work for machine operations in the field

RATE OF WORK	Time element

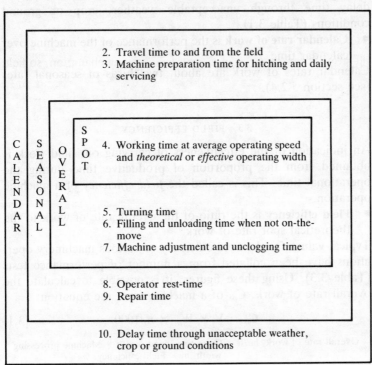

1. Machine preparation time before and after storage

2. Travel time to and from the field
3. Machine preparation time for hitching and daily servicing

4. Working time at average operating speed and *theoretical* or *effective* operating width

5. Turning time
6. Filling and unloading time when not on-the-move
7. Machine adjustment and unclogging time

8. Operator rest-time
9. Repair time

10. Delay time through unacceptable weather, crop or ground conditions

C A L E N D A R | S E A S O N A L | O V E R A L L | S P O T

Table 3.2 Average combine harvester performance using a 3 m width of header in barley

Theoretical spot rate of work	1.50 ha/h	—
Effective spot rate of work	1.35 ha/h	6.8 t/h
Overall rate of work	1.04 ha/h	5.2 t/h
Seasonal rate of work	0.92 ha/h	4.6 t/h
Field efficiency	69 %	

(*Source*: Elrick, 1982)

spot rate of work (Table 3.2). This clearly illustrates that the effective width of cut was only 90 per cent of the header width and further that the overall rate of work was only just over two thirds of the theoretical spot rate of work.

The *calendar rate of work* is occasionally used and includes the delay time through unacceptable weather, crop or ground conditions (Table 3.1).

- **Calendar rate of work** is the performance of the machine over calendar time.

Calendar rates of work are about two-thirds of seasonal rates (see section 5.2.4).

3.3 FIELD EFFICIENCY

An indication of the effectiveness of carrying out field work is obtained from the proportion of productive time during the operational time. This is called the *field efficiency* of a machine operation.

- **Field efficiency** is the ratio of the overall rate of work to the theoretical spot rate of work.

Typical values for the field efficiency of various machinery operations have been collated from a number of performance tests (Table 3.3). Using these figures, it is possible to calculate the overall rate of work, C_o, of a machine from the equation:

$$C_o = V \times W_m \times n_f/1000 \qquad \dots \text{[3.1]}$$

Overall rate of work, ha/h $\propto$ Operating speed, km/h $\times$ Machine processing width, m $\times$ Field efficiency, %

The maximum operating speed is controlled either by the operational requirement or by the comfort of the operator and the processing width of the machine is fixed, so it is essential to maintain a high field efficiency.

Field efficiency is achieved by ensuring that the unproductive time in machine operation is kept to a minimum. Fast speeds, wide machines, long fields, quick turns and rapid materials handling procedures all contribute to a high overall rate of work. The same list of factors, therefore, influences field efficiency:

- **theoretical spot rate of work;**
- **unused machine capacity;**
- **materials handling procedures;**

Table 3.3 Range in typical field efficiency values and operating speeds for field operations

Operation	Field efficiency, %	Operating speed, km/h
Tillage		
Mouldboard plough	75–85	5–10
Rotary cultivator	80–90	3–4
Disc harrows	80–90	6–12
Crop establishment		
Fertiliser broadcasting (inc. transport)	45–55	6–8
Combine drill	60–70	4–8
Potato planter	60–70	2–4
Sprayer	55–65	7–12
Harvesting		
Rotary mower	75–85	7–10
Forage harvester	60–70	4–6
Roll baler	70–80	5–8
Combine harvester	65–75	3–6
Potato harvester	65–75	2–4

○ **turning techniques;**
○ **fieldwork patterns;**
○ **field size and shape;**
○ **machine adjustments and blockages;**
○ **system limitations.**

Good management will always improve output but, just occasionally, this can be at the expense of a drop in field efficiency. For a more complete understanding of field efficiency, the factors which affect it are examined in more detail in the following sections.

3.4 THEORETICAL SPOT RATE OF WORK

Field efficiency decreases with an increase in the theoretical spot rate of work. This is because the increase in the theoretical spot rate of work has been obtained through using either a large machine, a faster speed, or a combination of both. Changing to a larger machine, at the same operating speed, allows more productive work to be accomplished in unit time but, as a result, the non-productive time represents a greater loss in potential production. For example, doubling the width of a stone and clod windrower from two rows to four rows only produced a 70 per cent increase in the overall rate of work – a worthwhile improvement but not directly proportional to the increase in size. A faster operating speed with the same width of machine again reduces field efficiency for the same reason as before. It is only under these circumstances that a lower field efficiency is associated with a higher overall rate of work; when the spot rate of work remains constant, any improvement in the field efficiency automatically improves the overall rate of work.

3.5 UNUSED MACHINE CAPACITY

In many field operations, machine bouts are overlapped to attain complete field coverage; utilising rather less than the full processing width of a harvester also ensures that no crop remains unharvested.

The extent of the overlapping, however, has a significant effect on field efficiency because it reduces the effective width of the machine by some 5 per cent (see section 3.2.1). The only exception to overlapping inefficiencies is with machines for crops grown in distinct rows, the design width and the effective width of a two-row potato planter being identical, for example.

Throughput capacity of a machine is also not always fully utilised. In forage harvesting, operating speed is seldom adequately increased to compensate for a very light crop. Gathering two swaths into one can achieve a dramatic improvement in the rate of work of the harvester, although the increase is partially offset by the additional machine operation.

3.6 MATERIALS HANDLING PROCEDURES

Materials handling is a major part of field work; seed, fertiliser and agrochemicals are distributed and forage, grain, straw, roots and vegetables are collected. Some of the machine filling and emptying procedures are carried out on-the-move but many involve stoppages. These stoppages may represent 20 per cent of the total operational time and the delays often can be halved by avoiding excessive manoeuvring, by using larger hoppers and by mechanical handling of stillages.

Crop spraying provides a good example of the various opportunities for improving machine output by combining better materials handling procedures with higher spot rates of work. Good management and adequate logistical support are vital to ensure rapid ground coverage during a short timespan which is restricted by both the stage of crop growth and the acceptable weather conditions for effective treatment. A typical system has three essential elements, the time taken to fill the tank, the time spent travelling to and from the filling point, and the time actually involved in the spraying operation. The relative importance of these time elements is affected by:

(a) refilling rate: e.g. change from a slow fill of 1 ℓ/s to a fast fill of 3 ℓ/s;

(b) location of filling point: e.g. change from farm supply to

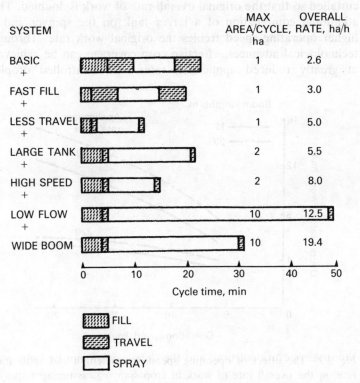

Fig. 3.1 The effect of cycle times on the performance of a crop spraying operation.

field bowser, reducing travel time from 5 minutes to 1 minute each way;

(c) tank size: e.g. change from 300 ℓ capacity to 600 ℓ capacity;
(d) spraying speed: e.g. change from 2 m/s to 4 m/s;
(e) application rate: e.g. change from 300 ℓ/ha to 60 ℓ/ha;
(f) boom length: e.g. change from 12 m to 20 m.

Considering a rectangular 12 ha field, 400 m by 300 m, and allowing 20 seconds for a headland turn, the effect of the various changes is graphically presented in Fig. 3.1. For the comparison, no account is taken of swath overlap, although it has been shown to average 6 per cent of boom width. In the basic system, the sprayer returns to the farm for refilling from a mains water supply. A fast fill from a high level tank gives some improvement but, with the use of a water bowser in the field travel time is also curtailed so that the original overall rate of work is doubled. The additional combination of a larger tank on the sprayer and a higher operating speed trebles the original work rate. Through technological advances, effective crop coverage can be achieved at greatly reduced application rates with controlled droplet

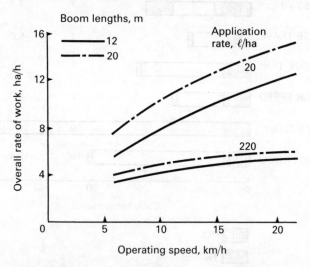

Fig. 3.2 The effect of operating speed, boom length and application rate on the overall rate of work of crop spraying, assuming a sprayer tank capacity of 600 ℓ and a rectangular, 12 ha field, 400 m by 300 m (after Nation, 1978).

application and with electrostatic deposition; wider booms can also be used through improvements in boom suspension systems. Thus, the adoption of all the system changes leads to nearly a ten-fold increase in work rate and, at the same time, reduces the proportional significance of the materials handling procedures because fewer materials are required.

Only a few of the possible combinations are shown in Fig. 3.1. Using a mathematical model, it is possible to investigate rapidly the likely effect on the overall rate of work of variations in any one or two of the range of parameters (Fig. 3.2).

3.7 TURNING TECHNIQUES

Turning time is the one non-productive element which is built into every field operation. Although the number of turns may be reduced by using wider machines and by creating larger fields, they can never be eliminated completely and correct execution, therefore, minimises delay.

Field work is organised either round the field in a *circuitous pattern* which involves a corner at the end of each side, or in a *parallel pattern* which involves full turns on each headland (see section 3.8 and Fig. 3.10). For a rectangular field, the main types of manoeuvre are:
○ **round corners;**
○ **square corners;**
○ **loop corners;**

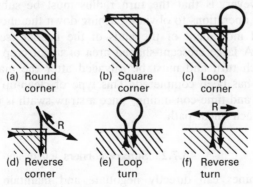

Fig. 3.3 Different types of manoeuvres.

○ **reverse corners;**
○ **loop turns;**
○ **reverse turns.**

These corners (or half turns) and full turns are shown in Fig. 3.3.

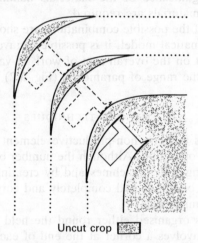

Uncut crop

Fig. 3.4 Round corners with crescent-shaped areas of uncut crop between each turn (adapted from Kepkay, 1972).

3.7.1 Round corners

This manoeuvre appears to be the most effective because the turn is accomplished during productive work (Fig. 3.4). The draw-back, however, is that the turn radius must be substantial in harvesting operations to obviate pushing down the uncut crop by the lateral movement of the ends of the header (see Fig. 3.5 point F). A large crescent-shaped area of uncut crop has to be left at each turn and must be salvaged after the main cutting operation has been completed. This type of clean-up operation is difficult and time-consuming, since a straw swath is interposed between each uncut path.

3.7.2 Square corners

Few machines can directly negotiate and maintain a square corner. For a combine harvester pivoting in its own width, the

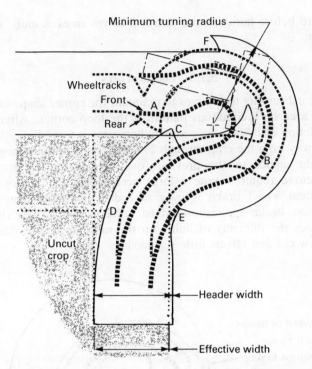

Fig. 3.5 Path of a combine harvester attempting to maintain a square corner.

minimum turning radius of the outer corner of the header is still 1.15 times the header width (Fig. 3.5). The sharpest turn involves two rapid steering reversals, firstly at point A to move the header away from the edge of the standing crop left behind so that, after cornering, the second sharp steering reversal at point B allows the header to sweep back into the crop without leaving any stalks uncut at point C. The corner of the header then follows a curved path, from C, through the crop but has still not straightened up by point D, so there will not be a square corner at that point on completion of the next circuit. Any attempt to effect a tighter sweep into the crop after negotiating the corner causes crop loss at point E because the locus of points for the rear of the header is then backwards and similar to that shown at point F. On the next circuit, the combine harvester has to proceed further forward before initiating the turn and leaves an even more acute corner.

3.7.3 Loop corners

The usual turning techniques to maintain the corner shape of the uncut area on a circuitous pattern is the loop corner. After the completion of its cut, the combine harvester is turned in a direction away from the crop through 270 degrees, starting its new cut at right angles to the previous one (Fig. 3.6). This turning manoeuvre requires quite a large area unless the appropriate individual wheel brake is used to reduce the loop diameter. Excessive brake application, whilst reducing the turning circle, increases the difficulty of lining up the machine before starting the new cut and effects little real saving.

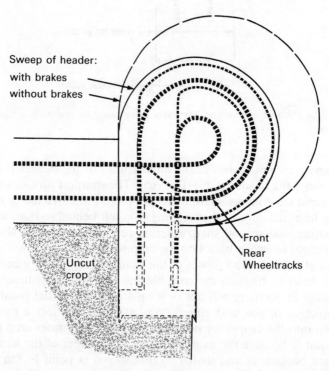

Fig. 3.6 A loop corner with and without brakes.

The loop corner is the quickest cornering manoeuvre with a mean observed time of 23 seconds, for both self-propelled and tractor-drawn machines over a wide range of operating speeds. This observation can be explained by approximating a loop corner to a complete turning circle. The turning time is then given by the equation:

$$t_{to} = 3600 \times 2 \times \pi \times r_m/(1000 \times V) = 23 \times r_m/V \ldots [3.2]$$

Turning time, s $\propto$ Mean turning radius, m/Operating speed, km/h

Substituting the observed turning time in the equation, the mean turning radius becomes numerically equal to the operating speed. The faster the speed, the larger the loop corner – space permitting – for the same turning time.

3.7.4 Reverse corners

This type of manoeuvre is often carried out inefficiently and without the benefit of applying individual wheel brakes. This is

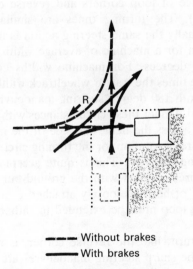

——— Without brakes
——— With brakes

Fig. 3.7 A reverse corner showing the difference between the correct technique with brakes and the commoner but slower option (adapted from Kepkay, 1972).

because the typical procedure is to drive the machine straight ahead to clear the corner, then to reverse on a curved path until it is aligned laterally with the new cut, and finally to proceed forward again in a straight line into the crop (Fig. 3.7).

When travelling in reverse, however, individual wheel brakes are much less effective, and steering is more difficult to judge and control. A better technique is to turn sharply into the corner, using the appropriate individual wheel brake, immediately after the header clears the standing crop. The machine is then reversed in a straight line, angled at 45 degrees to its original path to a position which permits another braked manoeuvre to effect a straight entry for the next swath. With self-propelled equipment, the reverse corner takes about 30 seconds, a third longer than the loop corner. The reverse corner with trailed machines is even slower at about 40 seconds, giving an average speed during the turn of 2 km/h.

3.7.5 Loop turns and reverse turns

For field work in a parallel pattern, loop turns and reverse turns are used in place of loop corners and reverse corners, respectively (Fig. 3.8). The turning times are similar to those for corners, as virtually the same steering action is involved.

The loop turn for a machine of average width requires rather more than 180 degrees. For machine widths equivalent to or more than three times the tractor wheeltrack width, the loop turn becomes a smooth 180 degree steering manoeuvre which can be quickly executed. Under these circumstances with a small overlap between passes across the field, one side of the machine extends beyond the centre of rotation for the turning circle and is pivoting backwards during the turn. This is quite acceptable for sprayer booms not in direct contact with the ground but could be disastrous with a set of harrows loosely attached to a main frame so that the turning loop must be extended to rather more than 180 degrees again.

The reverse turn, like the reverse corner, is most effectively excuted when the sharp steering manoeuvres are in the forward direction. For integrally mounted and for semi-mounted equipment, the turning time is similar to that for reverse corners with self-propelled machines. There is, however, additional time associ-

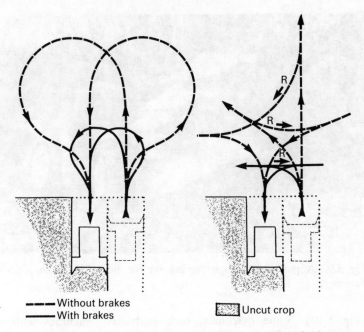

- - - - Without brakes
——— With brakes

▨▨▨ Uncut crop

Fig. 3.8 Loop turns and reverse turns for a parallel field pattern showing the typical paths without brakes and the more effective techniques (adapted from Kepkay, 1972).

ated with controlling machine attitude during the turn. Lifting and lowering equipment out of, and into, work is often smoothly incorporated into the turning technique. The introduction of push-pull tackle, such as front and rear-mounted ploughs, both reversible, involves slewing both units from one side of the tractor to the other as well as turning both ploughs over when in the centralised position (Fig. 3.9). This is feasible on-the-move but is a daunting mental task, with the result that the turning time gradually becomes more protracted and stabilises at about 60 seconds – double that for self-propelled machines.

3.8   FIELDWORK PATTERNS

Substantial improvements in field efficiency can accrue from choosing the most suitable pattern for field operations

Fig. 3.9 A push-pull plough turning on the headland (from *Arable Farming*).

(Fig. 3.10). Some machines, such as trailed machines with a laterally offset intake (e.g. mower) and conventional ploughs, are designed to operate in a clockwise direction with respect to the unprocessed area. Others, particularly, cultivators, front-mounted equipment and self-propelled machines, impose less restriction on the acceptability of the different procedures. Selection of the optimum fieldwork pattern is then dictated by other considerations, depending partly on field configuration (see section 3.9) and partly on minimising the amount of non-productive time spent on travel and turns. The fieldwork patterns can be separated into two groups, with variations according to turning technique:

○ **circuitous** (round and round) – travelling parallel to *all* the field boundaries, negotiating a corner at the end of each side;
○ **headland** – travelling parallel to *one* of the field boundaries, making a full turn on the headland after each pass across the field.

The circuitous pattern is easy to follow when starting from the field boundaries inwards. Where round corners prove difficult, in ploughing for example, turn strips on the field diagonals can be left uncultivated until the rest of the field is finished. It is more

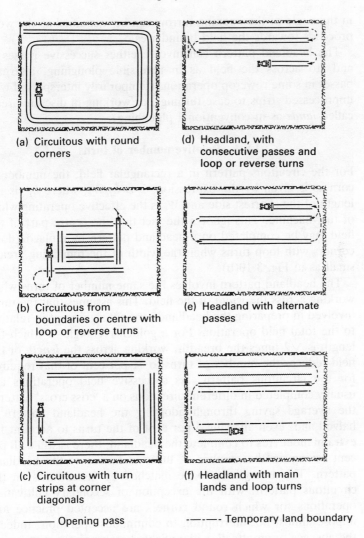

(a) Circuitous with round corners

(d) Headland, with consecutive passes and loop or reverse turns

(b) Circuitous from boundaries or centre with loop or reverse turns

(e) Headland with alternate passes

(c) Circuitous with turn strips at corner diagonals

(f) Headland with main lands and loop turns

——— Opening pass   –·–·– Temporary land boundary

Fig. 3.10 Typical fieldwork patterns.

difficult, however, to mark out a field in practice than it is in theory, particularly when starting from the centre of the field outwards. Undulations may obscure sight lines or vary the effective machine width on side lying land and relatively small errors in the angle of the opening furrow can be magnified as the work proceeds towards the field boundaries.

The headland pattern can involve either successive passes to and fro across the field as in reversible ploughing, alternate passes in some rowcrop operations (temporarily interspersed with unprocessed strips to ease turning), or working in discrete areas, called *lands* as in conventional ploughing.

3.8.1 Relative number of turns

For the circuitous pattern in a rectangular field, the number of corners is given by the expression, $2 \times b_f/W_e$, where b_f is the length of the shortest side and W_e is the effective operating width of the machine. This ignores the fact that the centre part of the field may be completed on a headland pattern by replacing loop corners with loop turns when the width of the remaining area is small as in Fig. 3.10(b).

The headland pattern involves the same number of turns when working across the *width* of the field. This assumes that the turns involved in preparing the headlands are not significant in relation to the total field operation. For a golden rectangle in which the length is $\sqrt{2}$ times the breadth, working across the *length* of the field reduces the number of turns to 71 per cent of that required for the circuitous pattern. As successive field operations are usually completed in different directions on a 'criss cross' pattern, the average saving through adopting the headland pattern is halved and, even allowing 5 per cent of the turns to account for extra manoeuvres to complete the headland, there is still a 10 per cent saving on the number of turns in favour of the headland pattern. This is unlikely to be matched by faster turns on the circuitous pattern, with the exception of secondary cultivation operations for which round corners are accepted practice and impose almost no interruption to continuous operation. Indeed, the absence of an effective identifiable break-point makes it very difficult to obtain comparative data on turn time for round corners.

3.8.2 Headland width

From an agronomic viewpoint, headlands are almost discard areas. There is reduced crop potential through greater soil compaction and through greater risk of weed or disease infestation close to the field boundaries. The insistence on narrow headlands to increase crop yield must be offset against the economic consequences of slower workrates. Inadequate width of headland is a major source of delay because a reverse turn must be substituted for the faster loop turns.

In general, the width of the headland should be 1½ to 2 times the combined length of the tractor and trailed machine or the overall width, whichever is the greater. The lower value is acceptable for fully mounted equipment and for very wide machines where the width of the wheeltrack is no more than one third of the total width.

3.8.3 Pattern comparison

The most important choice of fieldwork patterns is for ploughing using either a conventional plough which only throws the furrows to the right, or a reversible plough which has both right and left-handed bodies to permit consecutive passes.

For conventional ploughing, the field is set out in lands

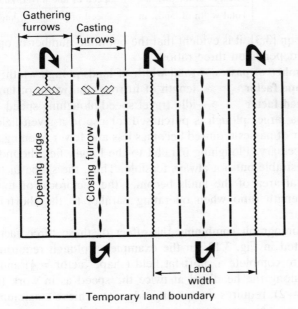

Fig. 3.11 Ploughing in lands to minimise idle running time.

(Fig. 3.11). An opening ridge is formed at the centreline of the first and subsequent alternate half lands. By ploughing along both sides of these opening ridges in a clockwise direction, the furrows are gathered, or lean towards, the mid-line, until the half land is completed. Thereafter, the intervening half lands are ploughed in the opposite direction to cast the furrows and leave an open furrow, or finish, equidistant between each pair of opening ridges. Either gathering or casting too large an area gives excessive wasted time in headland travel, but a greater number of lands requires more setting out time.

Making use of the method study procedure described in section 2.3.5, and mathematically analysed in Appendix A1, the optimum number of opening ridges for a rectangular field is given by the equation:

$$N_r = 0.5 + \left\{ \frac{1}{16} \times \frac{W_a}{W_p} \left[\frac{l_f}{W_a} \times \frac{V_i}{V_p} \right] \right\}^{0.5} \quad \ldots [3.3]$$

Number of opening ridges $\cong$ $\dfrac{\text{Total width of lands, m}}{\text{Width of plough, m}}$

$\div \left[\dfrac{\text{Length of furrow, m}}{\text{Total width of lands, m}} \times \dfrac{\text{Speed of idle travel, km/h}}{\text{Speed of ploughing, km/h}} \right]$

From eqn [3.3], it is evident that the optimum number of opening ridges depends on three ratios:

○ **number of passes** – total width of lands/plough width;
○ **shape factor** – length of furrow/total width of lands;
○ **speed factor** – idle travel speed/ploughing speed.

When a larger plough is purchased for use in a given field, the number of passes required becomes less and fewer opening ridges are necessary. Ploughing parallel to the longer field boundary is also desirable but not always feasible. Thus, the shape factor for the total area of the lands becomes the reciprocal of the field width/length ratio when operating parallel to the shorter field boundary.

A nomograph combining the effect of these three factors is illustrated in Fig. 3.12. In the example, a plough requiring 300 passes to complete a short fat field (shape factor = $\frac{1}{3}$) and travelling along the headland at twice the speed as in work (speed factor = 2), requires six opening ridges to achieve maximum field efficiency. This presents quite a daunting task for setting out the

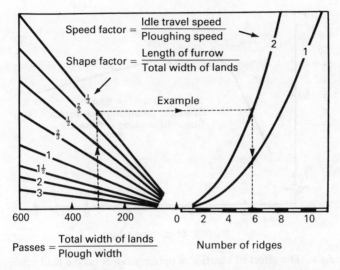

Fig. 3.12 Fieldwork pattern planner to calculate the number of opening ridges for maximum field efficiency with conventional ploughing.

field but ploughing as a long, narrow field of the same area would require only two opening ridges.

Considering a second combination of changing from a three-furrow semi-digger plough to a nine-furrow shallow plough, it is likely that the ploughing speed is substantially greater for the shallower depth of working so little increase would be expected in the idle travel speed (speed factor = 1). For a square field (shape factor = 1) requiring 600 passes of the small plough, the number of opening ridges is only reduced from five to four because of the additional effect of altering the speed factor as well as the number of passes.

This gives a rather better indication of the correct number of opening ridges required than the rule-of-thumb land width of 33 m for a two-furrow plough, 44 m for a three-furrow plough and so on. The fieldwork pattern planner is both quick to use and also visually demonstrates the relative merits of wider ploughs, long fields and fast travel on the headland in achieving a high field efficiency.

The effect of number of opening ridges on the field efficiency of conventional ploughing in a 12 ha field, 400 m by 300 m with

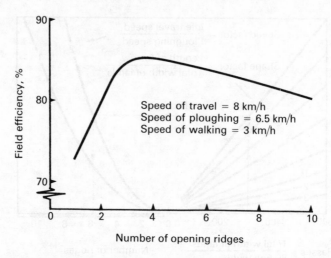

Fig. 3.13 The effect of number of opening ridges on the field efficiency of conventional ploughing in a 12 ha field, 400 m by 300 m with 5 m wide headlands, using a 1.2 m wide plough.

5 m wide headlands and sidelands, at a speed of 6.5 km/h with a 1.2 m wide, four-furrow plough is shown in Fig. 3.13. The walking speed and idle travel speeds are 3 km/h and 8 km/h, respectively, and the round corners to plough out the headlands and sidelands are assumed to take 0.005 h (18 seconds). As the optimum number of opening ridges is 3.5, the maximum field efficiency of 85 per cent for the complete operation, including setting out the field, is obtained for both three and four opening ridges. Too few opening ridges cause a greater reduction in field efficiency than too many.

With a reversible plough making consecutive passes across the field, there is no need to mark out sidelands unless the field shape is irregular. Assuming an average time of 0.01 h (36 seconds) for reverse turns on a 5 m wide headland, the field efficiency for the same width of plough and ploughing speed as previously is 88 per cent, only a marginal improvement on that for conventional ploughing.

The adoption of reversible ploughing has been encouraged for reasons far greater than field efficiency alone – convenience, less skill required for ploughing, and the ability to produce a level surface with less danger to the combine harvester table at harvest. For a medium sized, two-wheel drive tractor and a rear-mounted reversible plough, the second set of plough bodies provides extra weight to utilise a high drawbar power at maximum efficiency, especially on a heavy soil. On a light soil, however, the extra weight of a mounted reversible plough is of no advantage because drawbar power is restricted by the steering ability of the tractor. A long, heavy reversible plough tips the tractor up. In this situation, a mounted conventional plough can efficiently use 25 per cent more power than a mounted reversible plough. This represents the difference between a five-furrow conventional plough and a four-furrow reversible plough. The ability to pull an extra furrow with the conventional plough tips the maximum work rate in favour of ridges and furrows. Resurgence of interest in the land pattern is also strengthened by the trend towards giant tractors for which gangs of trailed equipment are necessary to utilise the power available and, at the same time, to maintain tractor stability and uniformity of depth over undulations.

3.9 FIELD SIZE AND SHAPE

Amalgamating small fields to give a larger area not only helps to reduce the number of turns and interfield travel, but also increases the effectiveness of larger machines and can permit higher operating speeds. For this reason, the effects of field area and of spot rate of work on the area capacity of machines are examined together.

When two small rectangular fields are combined to form one larger field of similar proportions (Fig. 3.14), the number of turns decreases theoretically to 71 per cent of the original number (see section 3.8.1). In practice, the machine width is such that the last pass in the smaller fields has to be partly overlapped but it might just as easily happen that the final pass of the field ended at the end of the field farthest from the gate. In either case, extra turns are involved to account for these practical considerations.

In a study of the economies of scale in farm mechanisation, it has been shown that the proportion of time spent on productive work increases substantially with size of field (Table 3.4). For the comparison, it was assumed that fields of a standard shape – off

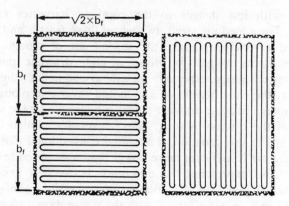

Fig. 3.14 The effect of amalgamating two small fields into one larger area of the same proportions.

the square with two sides not parallel – were cultivated by an implement with an effective width of 3 m and at an operating speed of 6 km/h. Each headland turn took 20 seconds and changing fields required 40 minutes. In the small, 2-ha fields, only 37 per cent of the time is spent cultivating whereas in a 80-ha field the productive time is doubled to 74 per cent.

As field size is increased, there is greater scope for wider machines. The effect of enlarging fields on the seasonal rates of

Table 3.4 Proportion of time spent on productive work in fields of increasing size

Field size, ha	Proportion of time on productive work, %				
	Productive work	Turns	Headlands	Changing fields	Contingencies
2	37	20	4	22	17
4	47	19	3	14	17
8	57	15	3	8	17
10	59	14	3	7	17
20	65	12	2	4	17
40	71	8	2	2	17
80	74	7	1	1	17

(*Source*: Sturrock *et al.* 1977)

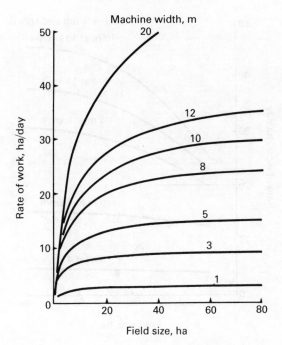

Fig. 3.15 The increase in work rate due to larger fields and wider machines at an operating speed of 6 km/h (adapted from Sturrock *et al.*, 1977).

work of six different widths of machine is shown in Fig. 3.15. An increase in field size from 10 ha to 20 ha enhances the work rate of a 1 m-wide implement, such as a plough, by 9 per cent; but with a 20 m-wide machine, such as a sprayer, the increase is 31 per cent. Most farmers would agree that 20 ha fields are an advantage for arable land. Further enlargement, whilst more controversial in terms of preservation of the countryside, still provides substantial gains in workrate.

The effect of enlarging fields is also greater with higher operating speeds. Although some field operations demand low forward speeds, many are limited more by driver comfort. With improved seat and cab suspension systems, high-powered tractors can be utilised more efficiently at higher operating speeds. The combined effect of field size, machine width and operating speed is shown in Fig. 3.16.

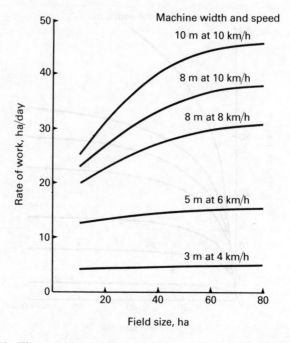

Fig. 3.16 The combined effect of larger fields, wider machines and higher speeds on rate of work (adapted from Sturrock *et al.*, 1977).

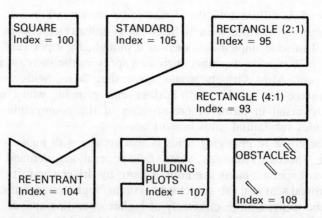

Fig. 3.17 Effect of shape on an index of comparative time to complete a 10 ha field, square field index = 100 (from Sturrock *et al.*, 1977).

An awkward shape can add appreciably to the time taken to cultivate a field. For a constant field area of 10 ha in the different shapes outlined in Fig. 3.17, estimates are given in Table 3.5 of the time required to cultivate (including turning, headlands and changing fields) with a 3 m-wide machine at an operating speed of 6 km/h. As already discussed in section 3.8.1, the rectangular fields require the least time, provided the operations are completed lengthwise, which may not always be the case. Thus, the square field is taken as a base line for the comparison of work rates.

Table 3.5 The effect of field shape on the time required to complete an area of 10 ha

Field shape	Time required, minute/ha	Index
Square	56.6	100
Rectangle (2:1)	54.0	95
Rectangle (4:1)	52.4	93
Standard – one side off square	59.5	105
Re-entrant side	59.1	104
Building plots	60.5	107
Obstacles	62.0	109

(*Source*: Sturrock *et al.* 1977)

Few fields, however, have parallel sides and the field shape adopted as standard has one side off the square which leaves a triangular area to be completed. In that area, the bouts become progressively shorter until the corner is reached. In comparison with a square field of the same size, the number of turns is increased (as well as being made more arduous) and the time is extended by 5 per cent. Although the triangular area could be avoided by operating across the field (because the other two sides are parallel), the number of turns in this specific case remains unaltered.

The re-entrant field with one side curving inwards leaves two triangles to be completed but the extra time involved is offset to some extent by the greater length of the field. For a shorter field of this general shape, it would be preferable to operate across the area and avoid the triangles. Building plots and other obstacles such as pylons, airfield flight path lights or isolated trees present a nuisance out of proportion to their size but must be tolerated as an accepted fact! The two building plots represent a sixth of the total rectangular area and the obstacles occupy 2 per cent but the extra time is 7 per cent and 9 per cent, respectively, compared with the square field.

As field size is increased, the loss of time through awkward field shapes becomes less significant. With very wide machines, however, considerable skill is required to negotiate obstacles without excessive overlap or missed areas.

3.10 MACHINE ADJUSTMENTS, BLOCKAGES AND SYSTEM LIMITATIONS

Whilst major breakdowns and repairs are excluded from the calculation of the overall rate of work, routine machine adjustments and the clearing of blockages are an accepted part of field operations. Potato harvesters, for example, are prone to blockages caused by stones jamming the soil separating webs and by haulm and weeds hindering crop flow. The delays clearing blockages of this nature represent from 5–10 per cent of the actual harvesting time (compared with 2–3 per cent for mechanical breakdowns and up to 15 per cent system limitations of inadequate transport).

There is, however, rather less information on time spent in machine adjustment. Often these adjustments are desirable because of changing soil or crop conditions and timely action can avoid excessive blockage. Maintaining a sharp cutting cylinder on a forage harvester not only prevents blockages but also increases the potential throughput quite markedly. Thus, the adjustment time can lead to an overall time saving rather than a delay.

The capacity of individual machines is frequently restricted by other limitations in the system. Few field operations are completely independent of other production operations, and involve not only transport but also processing and storage facilities at a different location. Combine harvesting may be delayed because of poor grain drier throughput and forage harvesting may be interrupted if the rate of handling material at the silo is inadequate. Correct matching of machinery in a system requires careful recording and analysis of all the individual activities more fully described in section 2.3.4.

3.11 BREAKDOWNS AND REPAIRS

Breakdowns are field delays caused by sudden failure of a component, with consequential repair time likely to exceed half an hour. The downtime which prevents the machine from working while weather and crop conditions would otherwise be suitable is included in the calculation of the seasonal rate of work.

A series of field surveys has indicated that mechanical reliability is not a major problem in harvesting operations. Potato harvesters lose only 2–3 per cent of harvesting time through breakdowns, beet harvesters 5–10 per cent and combine harvesters about 5 per cent. On average the lost time through combine harvester breakdowns only amounts to one extra hour per year per 100 ha, that is three hours per 100 ha for a three-year-old machine and six hours per 100 ha for a six-year-old machine.

3.12 SUMMARY

Machine performance is specified as an area capacity in hectares/hour or as a throughout capacity in tonnes/hour.

The spot rate of work is a measure of machine performance in productive work, whereas the overall rate of work includes routine interruptions for turning or product handling and the seasonal rate of work includes other delays which prevent the machine from working in otherwise suitable conditions.

The field efficiency provides an indication of the effectiveness of carrying out field operations by minimising the proportion of non-production work.

Overlapping passes reduce the effective width of a machine by 5 per cent.

Stoppages for materials handling procedures, representing up to 20 per cent of the total operational time, can be halved by improvements in machine design and in system organisation.

Turning time on the headland is shortest with a loop turn (approximately 23 seconds), whilst a reversing manoeuvre takes at least a third longer and may extend to over one minute when manipulating additional machine attitude controls.

A fieldwork pattern with headlands involves fewer turns than for the circuitous pattern. Wide headlands decrease turning time.

When operating in lands, the optimum area depends on three factors – the number of machine passes, the shape of the field and the ratio of idle travel speed to productive work speed – and can be identified from a fieldwork pattern planner.

Larger fields, preferably rectangular, not only reduce turning time but also increase the effectiveness of wider machines and permit higher operating speeds.

Blockages amount to 5–10 per cent of the actual harvesting time for root harvesting machinery but are not a major problem during combine harvesting. System limitations, such as lack of transport, could amount to 15 per cent of the actual harvesting time irrespective of crop and mechanical breakdowns are usually in the range from 2 to 5 per cent.

4

MACHINERY OPERATING COSTS

OUTLINE

Machinery purchase prices; Depreciation; Net interest; Vehicle excise duty and insurance; Shelter; Fuel and oil; Maintenance and repairs; Labour charges; Average operating costs; Discounted cash flows; Present annual costs of machine ownership; Investment grants; Tax allowances.

APOSTROPHE – BLACKMAIL AND SILVERMAIL

In the days when cattle were not only used as a guide to a man's personal wealth but also the main source of currency, the MacGregor clan founded a protection racket based on local black Highland cattle. The passing of coin as a means of payment was commonly known as 'silvermail' and the use of cattle as 'blackmail' from the Old Norse, mal, an agreement. The MacGregors levied blackmail against their neighbours in return for guaranteeing the safety of their cattle. If the offer was refused and payment withheld, the livestock in question would inevitably disappear. As soon as the premium was safely in the MacGregors' hands, these beasts would quickly reappear again.

(Source: Barrington, 1984)

4.1 INTRODUCTION

The operating costs for a machine are divided into two categories, *fixed costs* and *running costs*.

Fixed costs are dependent on the duration of ownership of a machine and include:

o **depreciation;**

○ **interest;**
○ **tax and insurance;**
○ **shelter.**

Running costs vary in proportion to the utilisation of a machine and comprise:

○ **fuel and oil;**
○ **repairs and maintenance;**
○ **labour.**

The distinction between fixed costs and running costs is not always clear cut. Whilst depreciation, or the loss in value of a machine with age, is stated as a fixed cost, this is only realistic under average conditions of operation. As the economic life of the equipment is reduced by heavy usage, part of the depreciation charge is dependent on utilisation. Conversely, repairs and maintenance is taken as a running cost but maintenance may still be required, even when the machine is little used. Thus, part of the running cost is linked to duration of ownership.

The labour charge is also taken as a running cost but this is only strictly correct for casual workers hired by the day. It may be convenient to apportion labour costs at an hourly rate: nevertheless, it must be recognised that wages represent a fixed annual commitment which can be amended only by careful forward planning and management. Despite these overlapping areas, the arbitrary division of the various financial charges into fixed costs and running costs is helpful in contrasting their overall effect on machinery operating costs. In the calculation of the *annual oper-*

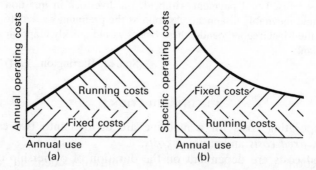

Fig. 4.1 The contrasting effect of fixed costs and running costs in the calculation of: (a) annual operating cost; (b) specific operating costs.

ating cost, the total fixed costs are constant per annum and the total running costs increase directly in proportion to annual use (Fig. 4.1(a)). Conversely, for the determination of the *specific operating cost* (that is the machinery charges per unit of use in terms of area covered, or material handled, or operational time), it is the *unit* running costs which remain constant, whereas the *unit* fixed costs decline because the standing charges are spread over increased annual use (Fig. 4.1(b)). In both approaches, the annual use of farm machinery is restricted because of the need for timely operations, without which severe losses in crop yield can accrue. As these penalties for untimely operations may be more important than all the machinery operating costs combined, they are considered separately.

4.2 PURCHASE PRICES

Many of the operating costs, either directly or indirectly, are linked to the list price of the machinery. An estimation of the capital cost of different sizes of equipment is useful when matching various tractor power outputs and machine capacities to provide the optimum machinery complement for a whole farm operation. A large price surcharge for a marginal increase in machine size has little market appeal because the extra cost would not be justified by the small improvement in output.

4.2.1 Marginal cost of extra machine capacity

In a highly competitive market, the list price of various categories of machinery is closely determined by the manufacturing costs for the design specification, plus a realistic profit margin. It is not surprising, therefore, to find that the prices of two-wheel drive tractors, all basically of the same design, are linearly affected by the engine power ratings:

$$PP_{2wd} = 2000 + 190 \times P_{max} \qquad \dots [4.1]$$

Purchase price of 2WD tractor, £ $\propto$ Maximum p.t.o. power, kW

A substantial change to the specification not only alters the production costs but also affects consumer demand and marketing strategy. The purchase prices of four-wheel drive tractors and for crawlers follow a similar pattern to those for two-

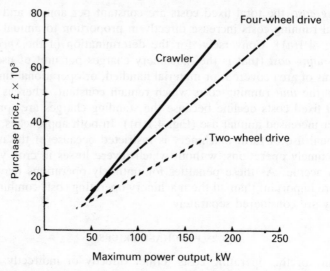

Fig. 4.2 The purchase prices of tractors related to power output.

wheel drive tractors but the price trends are increasingly affected by the engine power rating (Fig. 4.2).

Purchase price trends can be established for machinery in the same way. The major cost elements for a plough are the main frame per furrow (or implement width) and the number of plough bodies. These cost elements are linked in a linear equation with three coefficients, q, r and s:

$$PP_p = q + r \times W_f + s \times N_b \qquad \ldots [4.2a]$$

Purchase price of plough, £ ∝ Furrow width, m + Number of plough bodies

A change in frame design from a mounted design to either semi-mounted or trailed versions and from conventional to reversible specification jointly affects the independent coefficient and the machine width coefficient. The addition of accessories (e.g. automatic resetting of the plough leg after hitting an obstruction) affects the plough body coefficients only, the reversible plough having twice as many bodies as the same size of conventional plough.

For grain drills, the cost elements are the machine width per coulter and the number of coulters and, using different values of the coefficients, the purchase price is given by:

$$PP_d = q + r \times W_c + s \times N_c \qquad \ldots [4.2b]$$

Purchase price of grain drill, £ ∝ Coulter spacing, m + Number of coulters

The frame design (for mounted or trailed versions and for grain only or combined grain and fertiliser versions) and the coulter type (for prepared seedbeds or for direct drilling) influence the coefficients.

A slightly different form of the price equation is used for power driven cultivators, in that the price coefficient, r, is related to machine width rather than to the spacing between the tines:

$$PP_c = q + r \times W_m + s \times N_t \qquad \ldots [4.3]$$

Purchase price of cultivator, £ ∝ Machine width, n + Number of tines.

This change to machine width is necessary because tine or blade spacing on a rotary cultivator is related not only to distance but also to the angular position on the rotor.

The purchase price coefficients for various machine types are listed in Table 4.1.

Table 4.1 Values of price coefficients, q, r and s, for different types of machines

Machine	Purchase price coefficients		
	q	r	s
Tractors 1983			
Two-wheel drive	2 000	190	—
Four-wheel drive	−320	310	—
Crawler	−9 400	450	—
Ploughs 1983			
Conventional, mounted, fixed leg	−250	370	490
Reversible, mounted, fixed leg	−1 800	5 800	940
Conventional, semi-mounted, auto-reset	−20 000	58 000	740
Reversible, semi-mounted, auto-reset	−130 000	360 000	1 500
Drills 1984			
Conventional, mounted, grain only	5 800	38 000	160
Conventional, trailed, grain and fertiliser	−4 800	28 000	270
Direct, trailed, grain and fertiliser	−8 700	45 000	340
Cultivators (power driven) 1984			
Rotary, L blade	−400	350	31
Rotary, tine	−820	1 300	6
Harrow	−2 600	2 000	—

4.2.2   Tractor and machinery price indices

Purchase price data for only one year is severely limited in its usefulness, to the extent that reference to a current price guide is easier and quicker than applying the purchase price equations. Fortunately, the availability of official price indices simplify the updating of prices on a monthly or on an annual basis. The tractor price index and the farm machinery price index are listed in Table 4.2 for the past ten years. If the price of a new tractor was £PP_1 in 1984, and the tractor price indices for 1984 and 1986 are I_{t1} and I_{t2}, respectively, then the list price, £PP_2, of a similar new tractor in 1986 is:

$$PP_2 = PP_1 \times I_{t1}/I_{t2} \qquad \ldots [4.4]$$

$$1986 \text{ Purchase price, £} = 1984 \text{ Purchase price, £} \times \frac{1986 \text{ Tractor price index}}{1984 \text{ Tractor price index}}$$

The price estimated from the other purchase price equations can be projected forward in the same way.

Table 4.2   Price indices for tractors and for farm machinery for a 10-year period

Year	Price indices	
	Tractors	Machinery
1976	57.1	54.9
1977	71.3	66.2
1978	80.6	77.1
1979	89.4	88.3
1980	100.0	100.0
1981	108.5	104.9
1982	114.8	111.9
1983	124.9	114.2
1984	134.1	119.1
1985	143.8	126.5
1986	150.8	134.5

(*Source*: MAFF)

4.3   FIXED COSTS

4.3.1   Depreciation

Depreciation is the loss in value of a machine due to time and use. It is usually the largest cost component in machinery operations. Machines depreciate as a result of *age, wear and tear* and *obsolescence*.

Age A new machine is worth more than an old one, even though model changes have not significantly altered machine function.

Wear and Tear The more a machine is used, the greater the wear, reducing performance and reliability.

Obsolescence Novel machine concepts or higher throughputs may reduce the value of existing equipment.

There are several methods of calculating depreciation but the three most relevant are:

○ **straight-line depreciation;**
○ **declining balance depreciation;**
○ **decremental depreciation.**

Straight-line depreciation

With straight-line depreciation, an equal reduction in value is calculated for each year of machine ownership (Fig. 4.3). The *average annual depreciation* is obtained by deducting the resale value of the machine, *S*, from the purchase price, *PP*, and dividing by the period of ownership, *N*:

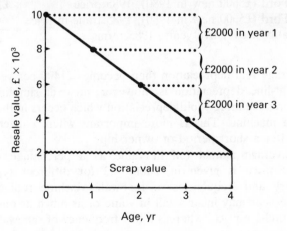

Fig. 4.3 Straight line depreciation.

$$D = (PP - S)/N \qquad \ldots [4.5]$$

Average annual depreciation, £ $= \dfrac{\text{Purchase price, £} - \text{Resale value, £}}{\text{Period of ownership, yr}}$

The purchase price relates to machines bought either new or second-hand, and the resale value, after a long period of ownership, may become the scrap value. During the early life of the machine, therefore, both the resale value and the period of ownership must be assumed.

The advantage of the straight-line depreciation method is that it is simple and straightforward. It is most suitable for estimating costs for the entire life of the machine. The annual depreciation charged can then be considered as the sum which must be set aside each year in order to replace the machine with an identical model at the end of the period of ownership. This, of course, assumes that the value of money remains the same and that the depreciation fund is not reinvested, that is, no inflation and no interest are considered.

Inflation can be accommodated, however, by calculating the depreciation in real terms for a machine which has been traded. It is obviously nonsense to say that a tractor has only cost £3000 over five years just because it has cost £8300 at 1980 values and is now (1985) worth £5300. The prices need converting to present-day values:

New Ford (£5600 new in 1980), 1985 price	£12 300
1980 Ford (£5600 new in 1980), 1985 trade-in price	5 300
Capital cost over five years, 1985 terms	£7 000

The average real depreciation then becomes £1400 per year.

Straight-line depreciation is, however, an over simplification and ignores the more rapid depreciation which occurs in the early life of a machine. This is more important when a machine is traded after a short period of ownership.

The average annual fall in value as a percentage of the purchase price is given in Table 4.3 for different types of machinery and frequencies of renewal. Machine replacement after one year may incur a fall in value of as much as one third of the purchase price, whereas a low frequency of renewal every eight years gives an average annual fall in value of only 10 per cent.

Table 4.3 Average annual fall in machinery values as a percentage of purchase price

Frequency of renewal, year	Average annual fall in machinery value, % purchase price		
	Complex equipment with high depreciation rate, e.g. potato harvesters, mobile pea viners, etc.	Established machines with many moving parts, e.g. tractors, combine harvesters, forage harvesters, balers	Simple equipment with few moving parts, e.g. ploughs, trailers
1	34	26	19
2	24½	19½	14½
3	20*	16½*	12½
4	17½†	14½	11½
5	15‡	13†	10½*
6	13½	12	9½
7	12	11	9
8	11	10‡	8½†
9	(10)	9	8
10	(9½)	8½	7½‡

* Typical frequency of renewal with heavy use.
† Typical frequency of renewal with average use.
‡ Typical frequency of renewal with light use.
(*Source*: Culpin, 1975)

Declining balance depreciation

With declining balance depreciation, a machine depreciates by a different amount each year, but the decimal rate of annual depreciation remains constant. The annual depreciation cost for any year is obtained by multiplying the written-down value of a machine at the beginning of the year by the decimal rate of annual depreciation, d (Fig. 4.4). The resale value of the 'n' year old machine is:

$$S = PP \times (1 - d)^n \qquad \dots [4.6]$$

Resale value, £ = Initial purchase price, £
$\times$ (1-Decimal rate of depreciation, dim.)$^{(\text{Age, yr, exponent})}$

The need to use an exponent makes the method more complex and tedious. It better reflects, however, the actual value of a machine at any age and is more useful for calculating the value of assets in a balance sheet.

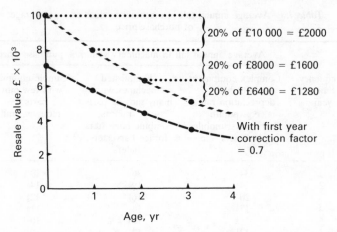

Fig. 4.4 Declining balance depreciation showing the additional effect of a first year correction factor.

Even so, the method does not account for first year depreciation which tends to be at a considerably higher rate than in later years. By introducing a first year correction factor to the resale value calculation, the accuracy of the declining-balance method can be improved still further. The resale value for an '*n*' year old machine as a decimal proportion of the initial purchase price is:

$$S/PP = C_f \times (d_f)^n \qquad \ldots [4.7]$$

$$\frac{\text{Current resale value, £}}{\text{Current purchase price, £}} = \begin{bmatrix} \text{First year} \\ \text{correction} \\ \text{factor, dim.} \end{bmatrix} \times \begin{bmatrix} \text{Annual} \\ \text{depreciation} \\ \text{factor, dim.} \end{bmatrix}^{\text{(Age, yr, exponent)}}$$

American machinery is classified into four groups for estimating the resale value and the relevant factors are listed in Table 4.4.

The use of these correction factors is not really the complete answer because the improved approximation to the actual resale value in later years is offset by the error in the resale value of a brand new machine.

Decremental depreciation

In practice, the ratio of the resale value to the initial purchase price is an exponential function. The actual data in Table 4.5 for

Table 4.4 Resale value factors for various agricultural machines

Machine	First year correction factor (C_f)	Annual depreciation factor (d_f)	Rate of annual depreciation, %
Tractor (2WD, 4WD, crawler); stationary power unit	0.68	0.920	8
Combine harverster (p.t.o., self-prop.); crop drier (high temp.); forage wagon; swather (self-prop.)	0.64	0.885	11.5
Feed wagon; fertiliser equipment; flail harverster; fore end loader; hay conditioner; land plane; manure spreader; mower; pick-up truck; potato harvester; scraper; seeding equipment; side delivery rake; sprayer (mounted); sugar beet harvester; tillage equipment	0.60	0.885	11.5
Baler (p.t.o., self-prop.); ensilage blower; forage harvester (trailed, self-prop.); sprayer (self-prop.)	0.56	0.885	11.5

(*Source*: ASAE, 1982)

six popular models of two wheel drive tractors exhibit very close agreement to a single declining curve with the equation:

$$S/PP = e^{(-0.2 \times n + 0.008 \times n^2)} \qquad \ldots [4.8]$$

$$\frac{\text{Current resale value, £}}{\text{Current purchase price, £}} = \exp \text{ (Age, yr, related function)}$$

where e is the base of natural logarithms and n is the depreciation age of the tractor. The advantage of this relation is that the resale value of the tractor when new, at age zero, is identical with the current purchase price (Fig. 4.5).

This resale value assumes an average annual use of 1000 h/yr. Heavier or lighter use influences the condition of the tractor and its resale value. An indication of the *operational age* of a tractor

Table 4.5 The ratio of current resale value to current purchase price for various two-wheel drive tractors

	Current resale value/Current purchase price					
Age	MF 135	JD 1120	Ford 3000	MF 165	Ford 4000	IH 574
2	—	—	—	0.68	—	—
3	—	0.49	0.52	0.60	0.52	0.43
4	0.46	0.40	0.43	0.54	0.42	0.34
5	0.42	0.34	0.37	0.43	0.35	0.29
6	0.38	0.29	0.33	0.37	0.32	0.25
7	0.34	0.25	0.29	0.33	0.29	0.23
8	0.31	0.22	0.28	0.30	0.26	—
9	0.29	0.19	0.26	0.27	0.25	—
10	0.28	0.17	—	0.26	—	—
11	0.25	—	—	—	—	—
12	0.24	—	—	—	—	—

(*Source*: Kirkpatrick, 1979)

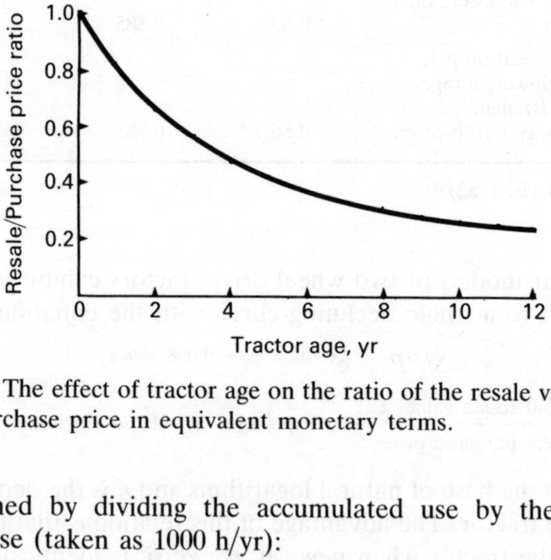

Fig. 4.5 The effect of tractor age on the ratio of the resale value to the initial purchase price in equivalent monetary terms.

is obtained by dividing the accumulated use by the average annual use (taken as 1000 h/yr):

$$\text{Operational age, yr} = \frac{\text{Accumulated use, h}}{\text{Average annual use, h/yr}}$$

For depreciation purposes, the *mean* of the actual age and the operational age provides a realistic depreciation age for assessing resale value:

$$\text{Depreciation age, yr} = \tfrac{1}{2} \times (\text{Actual age, yr} + \text{Operational age, yr})$$

because it accounts for the effect of ageing as well as utilisation. For example, a tractor sold after four years at 1500 h/yr is equivalent to an operational age of six years and the mean value or depreciation age is then five years. Similarly, a lightly used tractor of the same age completing only 500 hours annually would have an operational age of two years, so the depreciation age is three years.

4.3.2 Interest

In calculating the cost of owning and operating individual machines, there must be a charge for the interest on the capital which is invested in the equipment. Either the capital is borrowed and interest has to be paid, or the capital is already owned and has an opportunity cost which is the highest return that the capital could earn from an alternative investment. The opportunity cost of capital is by no means easy to define but a realistic allowance can be calculated from the earning power in a safe investment such as a Building Society.

Inflation and tax allowances make the real cost of borrowing much lower than indicated by overdraft interest rates, so it is much more appropriate to use investment interest rates. After one year, for example, the total repayment for a loan from the farmer's own capital of £100 at an annual rate of investment interest of 15 per cent is £100 + £15. If inflation is at an annual rate of 12 per cent, the purchasing power of the original £100 is equivalent to £100 + £12 at the end of the year. Thus, the real interest rate, using identical monetary values, is the total repayment related to the purchasing power of the loan. Expressed mathematically, the real interest rate, i_r, is:

$$i_r = \left[\frac{(1 + i_i/100)}{(1 + j/100)} - 1 \right] \times 100 \qquad \ldots [4.9]$$

$$\text{Real interest rate, \%} = \left[\frac{(1 + \text{Investment interest rate, \%}/100)}{(1 + \text{Inflation rate, \%}/100)} \right] - 1 \times 100$$

where i_i is the investment interest rate and j is the inflation rate. So in the example,

$$i_r = \left[\frac{(1 + 0.15)}{(1 + 0.12)} - 1\right] \times 100 = 2.68\%$$

Instead of fully representing the effect of inflation, a simple procedure is to remove the effect of inflation by using the concept of net interest rate. Net interest rate is the full rate of interest less the rate of inflation:

$$i_n = i_i - j \qquad \qquad \text{... [4.10]}$$

Net interest rate, % = Investment interest rate, % − Inflation rate, %

For example, the net interest rate is 3 per cent when the overdraft interest rate is 15 per cent and the inflation rate is 12 per cent. It has averaged about 3 per cent over the last 20 years or so and has not been subject to the same wild fluctuations as either actual interest or inflation.

When straight-line depreciation is being used, the *average investment* over the life of the machine is half the sum of the purchase price and resale value (Fig. 4.6). The *interest charge* is obtained from the average investment over the whole period of ownership multiplied by the net rate of interest:

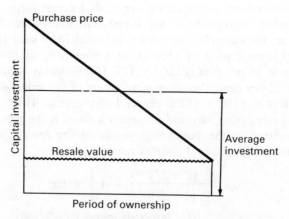

Fig. 4.6 Average investment when straight-line depreciation is used.

$$I = i_n \times (PP + S)/200 \qquad \ldots [4.11]$$

Interest charge, £ = Net interest rate, % × (Purchase price, £
+ Resale value, £)/200

The net rate of interest is rather higher than the real interest rate
but the effect is counter-balanced by two factors. Firstly, the
average investment is under-estimated because of lower depreciation after the first year (see Fig. 4.4). Secondly, the true
interest charged is increased because of the effect of compound
interest. The compounding surcharge has less significance over
a short economic life of, say, five years for machinery compared
with 20 years for buildings, and for lower rates of interest such
as those used for the net interest calculation. The overall result
is a simple, accurate assessment of interest charges for machinery
operating costs.

Over longer life spans of equipment and at higher rates of
interest, the appropriate interest charges begin to rise substantially by compounding. When repaying a loan by equal instalments, the initial repayments (or amortisation charges) are
largely repayment of interest whilst the final repayments are
mainly payment of capital (Fig. 4.7). Amortisation charges for
different rates of interest and loan periods are given in Table 4.6.

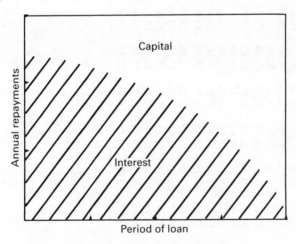

Fig. 4.7 The changing ratio of interest and capital repayment in amortisation charges over the duration of a loan.

Table 4.6   Amortisation charges and average interest payments per annum per £100

Term, year	Interest rate, %														
	4	5	6	7	8	9	10	11	12	13	14	15	16	18	20
I Amortisation charges															
2	53.01	53.78	54.54	55.30	56.07	56.84	57.61	58.39	59.16	59.94	60.72	61.51	62.29	63.87	65.45
3	36.03	36.72	37.41	38.10	38.80	39.50	40.21	40.92	41.63	42.35	43.07	43.79	44.52	45.99	47.47
4	27.54	28.20	28.85	29.52	30.19	30.86	31.54	32.23	32.92	33.61	34.32	35.02	35.73	37.17	38.62
5	22.46	23.09	23.73	24.38	25.04	25.70	26.37	27.05	27.74	28.43	29.12	29.83	30.54	31.97	33.43
6	19.07	19.70	20.33	20.97	21.63	22.29	22.96	23.63	24.32	25.01	25.71	26.42	27.13	28.59	30.07
7	16.66	17.28	17.91	18.55	19.20	19.86	20.54	21.22	21.91	22.61	23.31	24.03	24.76	26.23	27.74
8	14.85	15.47	16.10	16.74	17.40	18.06	18.74	19.43	20.13	20.83	21.55	22.28	23.02	24.52	26.06
9	13.44	14.06	14.70	15.34	16.00	16.67	17.36	18.06	18.76	19.48	20.21	20.95	21.70	23.23	24.80
10	12.32	12.95	13.58	14.23	14.90	15.58	16.27	16.98	17.96	18.42	19.17	19.92	20.69	22.25	23.85
11	11.41	12.03	12.67	13.33	14.00	14.69	15.39	16.11	16.84	17.58	18.33	19.10	19.88	21.47	13.11
12	10.65	11.28	11.92	12.59	13.26	13.96	14.67	15.40	16.14	16.89	17.66	18.44	19.24	20.86	22.52
13	10.01	10.64	11.29	11.96	12.65	13.35	14.07	14.81	15.56	16.33	17.11	17.91	18.71	20.36	22.06
14	9.46	10.10	10.75	11.43	12.12	12.84	13.57	14.32	15.08	15.86	16.66	17.46	18.28	19.96	21.68
15	8.99	9.63	10.29	10.97	11.68	12.40	13.14	13.90	14.68	15.47	16.28	17.10	17.93	19.64	21.38
20	7.35	8.02	8.71	9.43	10.18	10.95	11.74	12.55	13.38	14.23	15.09	15.97	16.86	18.62	20.53
II Average interest payments															
2	3.0	3.8	4.5	5.3	6.1	6.8	7.6	8.4	9.2	9.9	10.7	11.5	12.3	13.9	15.5
3	2.7	3.4	4.1	4.8	5.5	6.2	6.9	7.6	8.3	9.0	9.7	10.5	11.2	12.7	14.1
4	2.5	3.2	3.9	4.5	5.2	5.9	6.5	7.2	7.9	8.6	9.3	10.0	10.7	12.2	13.6
5	2.5	3.1	3.7	4.4	5.0	5.7	6.4	7.1	7.7	8.4	9.1	9.8	10.5	12.0	13.4
6	2.4	3.0	3.7	4.3	4.9	5.6	6.3	7.0	7.6	8.3	9.0	9.8	10.4	11.9	13.4
8	2.4	3.0	3.6	4.2	4.9	5.6	6.3	6.9	7.7	8.3	9.1	9.8	10.5	12.0	13.6
10	2.3	3.0	3.6	4.2	4.9	5.6	6.3	7.0	7.7	8.4	9.2	10.0	10.7	12.3	13.9
12	2.3	3.0	3.6	4.3	5.0	5.7	6.4	7.1	7.9	8.6	9.4	10.2	10.9	12.6	14.3
15	2.3	3.0	3.6	4.3	5.0	5.7	6.5	7.2	8.0	8.8	9.6	10.4	11.2	12.9	14.7
20	2.4	3.0	3.7	4.4	5.2	6.0	6.7	7.6	8.4	9.2	10.1	11.0	11.8	13.7	15.5

By deducting the average annual capital repayment per annum from the total annual repayment, the average interest payment is obtained. For example, at a rate of interest of 20 per cent over 15 years the average interest payment in 14.7 per cent on the total investment which gives a total interest charge 50 per cent higher than by the simple calculation approach.

4.3.3 Vehicle excise duty and insurance

The registration of agricultural tractors and self-propelled machinery for use on the public highway involves an annual payment of vehicle excise duty at a nominal rate of £15, provided that the governed engine speed of the vehicle does not permit a road speed in excess of 32 km/h. Otherwise, the tractor must be licensed as a haulage vehicle which attracts vehicle excise duty of £400/year and must use ordinary, non-rebated diesel fuel.

Insurance for tractors and self-propelled machines covers third party liability, fire and theft. In many cases, the insurance cover also includes free cover for attached trailers and for accidental damage to trailed machinery when attached to the tractor, including internal damage to the machine caused by the ingress of foreign objects. The premium is based on a minimum charge plus a sliding scale of rates per £100 of the sum insured and is typically about 1 per cent of market value. When equipment is detached, it may be covered by Fire and Farming Stock Insurance at a lower premium of 0.25 per cent of market value.

4.3.4 Shelter

Although the financial benefit of shelter for farm equipment is difficult to evaluate, most complex machinery is stored under cover for at least part of the year. Apart from facilitating maintenance and repair, a tidy machinery shed demonstrates a managerial commitment to good machine care. A purpose-built machinery store requires enclosed workshop facilities and garaging for self-propelled equipment. More robust implements are usually stored in an adjacent lean-to shed with open sides. Depreciating this type of structure over an economic life of 20 years generates an annual cost of 12 per cent of the initial purchase price of the machinery store. The shelter charge is then between 0.5 per cent and 1 per cent of the initial purchase price of the machinery stored inside the building.

The shelter, however, need not be a purpose-built structure. Tractors may be housed in old barns or cart-sheds, long since written-off and unsuitable for alternative uses. They may also be kept in umbrella buildings designed primarily as crop stores or in feed passages in cattle sheds. Under such circumstances, the shelter charge is likely to be no more than 0.2 per cent of the purchase price of the machinery.

4.4 RUNNING COSTS

4.4.1 Fuel and oil

The fuel consumption of a tractor is governed by the amount of energy demanded at the drawbar or through the power take-off. The net energy requirements for typical farm operations are shown in Table 4.7. One kilowatt of power delivered for one hour is one kilowatt-hour of energy or 3.6 megajoules:

$$1 \text{ kW} \times 1 \text{ h} = 1 \text{ kJ/s} \times 3600 \text{ s} = 3600 \text{ kJ} = 3.6 \text{ MJ}$$

In order to relate these net energy requirements to the tractor fuel consumption, it is necessary to account for the efficiency of the power transmission system, the tractive efficiency, and the loading on the engine. Typical fuel requirements for individual field operations are given in Table 4.8.

Table 4.7 Net energy requirement at the tractor drawbar or through the p.t.o. for typical farm operations

Operation	Draught power, kW	P.t.o. power, kW	Net energy requirement
Ploughing (3 furrows, 20 cm deep)	32	—	222 MJ/ha
(7 furrows, 10 cm deep)	38	—	108 MJ/ha
Rotary cultivation	4	38	82 MJ/ha
Grain drilling (15 row)	9	—	13 MJ/ha
Direct drilling (15 row)	14	—	22 MJ/ha
Forage harvesting	8	28	10 MJ/t
Potato harvesting	15	5	280 MJ/ha

Table 4.8 Fuel requirements for individual field operations

Operation	Fuel consumption, ℓ/ha
Subsoiling	15
Ploughing	21
Heavy cultivation	13
Light cultivation	8
Rotary cultivation	13
Fertiliser distribution	3
Grain drilling	4
Rolling	4
Potato planting	8
Mowing, tedding, baling	3
Forage harvesting	15
Spraying	1
Combine harvesting	11
Potato harvesting	21

Tractors operate throughout the year on a range of tasks varying from heavy duty work such as ploughing or forage harvesting to light chores. Even for ploughing, the fuel consump-

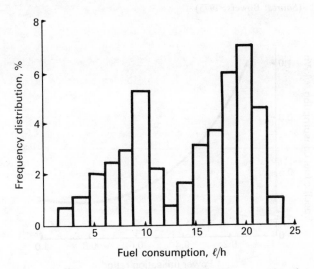

Fig. 4.8 Fuel consumption varies considerably over the duty cycle, even for heavy work such as ploughing (after Palmer, 1984).

tion of an individual tractor varies considerably over the duty cycle (Fig. 4.8) and the average fuel consumption is only two thirds of the fuel consumption for peak power. From a study in the United States, the percentage of time that a tractor operates at each power level is shown in Table 4.9. The *average engine loading* throughout the year is 55 per cent of the maximum power take-off (p.t.o.) power of the tractor.

Fuel efficiency varies with engine loading and reaches a maximum at about 90 per cent of maximum power. The fuel consumption per unit of power output or *specific fuel consumption* is shown for different diesel engine loadings in Fig. 4.9. The

Table 4.9 Percentage of time tractors operate at various engine loadings

Engine loading, % of max power	Operating time, % of annual use
Over 80	16.8
60–80	23.9
40–60	22.6
20–40	17.5
Under 20	19.2

(*Source*: Bowers, 1975)

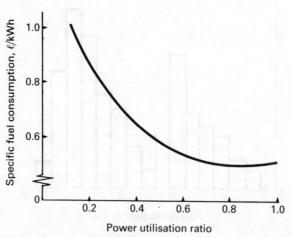

Fig. 4.9 Specific fuel consumption for a diesel engine operating at various power utilisation ratios.

specific fuel consumption, fc, for a diesel engine is related to the power utilisation ratio, R_u, by the expression:

$$fc = 2.64 \times R_u + 3.91 - 0.2 \sqrt{(738 \times R_u + 173)} \qquad \ldots [4.12]$$

$$\text{Specific fuel consumption, } \ell/\text{kWh} \propto \frac{\text{Equivalent p.t.o. power requirement, kW}}{\text{Maximum p.t.o. power, kW}}$$

In the power utilisation ratio, the equivalent p.t.o. power requirement is a measure of the engine power output required for either p.t.o. or for draught power.

The fuel consumption of self-propelled machinery is usually based on the requirement per hectare (Table 4.8).

Oil consumption is defined as the volume of engine crankcase oil per hour replaced at the recommended change intervals. The oil consumption, oc, is related to the rated engine power P_{max}:

$$oc = 0.021\ 69 + 0.000\ 59 \times P_{max} \qquad \ldots [4.13]$$

$$\text{Oil consumption, } \ell/\text{h} \propto \text{Rated engine power, kW}$$

This gives an oil consumption of 1 litre for every 300 litres of fuel for 25 kW tractors gradually decreasing to almost half that for 150 kW tractors (Table 4.10).

Table 4.10 Fuel and oil consumption for diesel engines of various power outputs

Max power output p.t.o. kW	Fuel consumption, ℓ/h		Oil comsumption, ℓ/100 h
	at max power	at 55% max power	
25	13	8	4
50	26	15	5
75	38	23	7
100	51	30	8
125	64	38	10
150	77	45	11

Assuming that diesel fuel price, p_f, is £0.18/litre and lubricating oil price, p_o, is £2/litre, the oil consumption costs, OC, represent about 5 per cent of the fuel consumption costs, FC.

$$FC + OC = fc \times P \times p_f + oc \times p_o \qquad \ldots [4.14]$$

Fuel cost, £/h + Oil cost, £/h = Specific fuel consumption, ℓ/kWh
× Equivalent p.t.o. power required, kW × Fuel price, £/ℓ
+ Oil consumption, ℓ/h × Oil price, £/ℓ

4.4.2 Maintenance and repairs

Maintenance and repairs are essential in an effort to guarantee a high standard of machine performance and reliability. Reliability is a measure of the confidence which can be placed on a machine to complete a planned duty cycle without a component failure. Over the life of a machine, components become worn. Excessive wear adversely affects output and increases the likelihood of random failures. This introduces a greater financial risk by prolonging a critical operation either through a slower rate of work or through an unforeseen breakdown.

Workshop care of machinery is of four types: *routine replacement of wearing parts, repair of accidental damage, repairs through operator neglect* and *routine overhauls*.

Routine replacement of wearing parts

Routine replacement of tines, discs, shares and mouldboards is required on tillage equipment; shearbars, flails and knives on harvesters; tyres and batteries on tractors. Although the wear rate is largely governed by soil characteristics, by crop flow, by engine hours or by natural deterioration, good maintenance can extend the serviceable life.

Repair of accidental damage

By their very nature, accidents can occur with any operator but carelessness, rushing to finish a job, and excessive overtime are the most likely causes of costly damage. Higher speeds coupled with a lapse of concentration can lead to extensive damage to a machine; for example, by hitting a gate post or by picking up tramp metal. The damage can involve twisting a main frame, cracking an axle housing, throwing a chopper mechanism out of balance or other parts which are both difficult to replace and hard to procure from the local dealer network because they are not standard stock items. Provided there is sound management to ensure that neither the operator nor the machine is overstretched, competence of the operator is the key to minimising accidental damage.

Repairs through operator neglect

Neglecting regular maintenance or minor repairs can be expensive. Engine oil starvation leads to overheating and even bearing failure, all for the sake of a timely oil check. Tyres are quickly ruined by underinflation. Slippage on slack running chains can cause timing errors and consequential damage to major components. Instead of a quick tensioning adjustment, a major overhaul is required. Good out-of-season maintenance together with a rigid daily inspection and servicing schedule go a long way towards ensuring trouble-free operation of equipment.

Routine overhauls

Machines are overhauled to replace worn or defective parts before total failure occurs. The service life of various groups of components is reasonably predictable; for example, clutches, drive belts, harvester webs, etc. Routine overhauls can be organised in advance and undertaken during slack periods when machine downtime is unimportant compared with the potential benefits from restoring the performance to the original level.

Machinery repair costs consist of the full financial outlay for parts and labour either at the dealers or on the farm. These charges are in themselves non-standard; labour costs for main dealerships with well equipped workshops are considerably higher than at a local garage, reflecting the different level of overheads. In addition, the differences in operator care, in soils, in crops and in climate produce a broad range of repair costs for similar machines. Ideally, an accurate record of repair and maintenance costs should be kept for each machine on the farm. As a guide, however, average repair costs are influenced by the size of the machine, as reflected by its price, and the amount of use. For any category of machines, there is an index of accumulated repair costs which is the ratio of the accumulated repair and maintenance costs, AR, to the initial purchase price. This repair cost index is given by a logarithmic expression based on the

Table 4.11 Values of the repair constant and repair exponent used in the calculation of accumulated repair costs for various types of machine

Machine type	Av field speed, km/h		Estimated life, h	Total life repairs, % of list price	Accumulated repair cost index	
	Typical	Range			Repair constant	Repair exponent
Tractors & Transport						
Two-wheel drive			10 000	120	0.012	2.0
Four-wheel drive & crawler			10 000	100	0.010	2.0
Trailer			3 000	80	0.19	1.3
Tillage						
Mouldboard plough	7.0	5.0–10.0	2 000	150	0.43	1.8
Heavy-duty disc	7.0	5.5–10.0	2 000	60	0.18	1.7
Tandem disc harrow	6.5	5.0–10.0	2 000	60	0.18	1.7
Chisel plough	7.0	6.5–10.5	2 000	100	0.38	1.4
Field cultivator	9.0	5.0–13.0	2 000	80	0.30	1.4
Spring tooth harrow	9.0	5.0–10.0	2 000	80	0.30	1.4
Roller-packer	10.0	7.0–12.0	2 000	40	0.16	1.3
Rotary hoe	11.0	8.0–16.0	2 000	60	0.23	1.4
Rowcrop cultivator	5.5	4.0 – 8.0	2 000	100	0.22	2.2
Rotary cultivator	5.0	2.0 – 7.0	1 500	80	0.36	2.0

Establishment						
Fertiliser spreader	1.3	0.95	120	1 200	7.0	5.0–8.0
Grain drill	2.1	0.54	80	1 200	6.5	4.0–10.0
Crop sprayer	1.3	0.41	70	1 500	10.5	5.0–11.5
Harvesting						
Combine harvester:						
trailed	2.3	0.18	90	2 000	5.0	3.0–6.5
self-propelled	2.1	0.12	50	2 000	5.0	3.0–6.5
Mower	1.7	0.46	150	2 000	8.0	6.5–11.0
Mower conditioner	1.6	0.26	80	2 000	7.0	5.0–10.0
Side delivery rake	1.4	0.38	100	2 000	7.0	6.5–8.0
Baler	1.8	0.23	80	2 000	5.5	4.0–8.0
Big bale baler	1.8	0.23	80	2 000	5.5	5.0–8.0
Forage harvester:						
trailed	1.8	0.23	80	2 000	4.0	2.5–4.0
self-propelled	1.8	0.12	60	2 500	5.0	2.5–10.0
Forage blower	1.8	0.14	50	2 000		
Sugar beet harvester	1.4	0.19	70	2 500	5.0	4.0–8.0
Potato harvester	1.4	0.19	70	2 500	3.0	2.5–6.5

(*Source:* ASAE, 1986)

accumulated use of the equipment, U:

$$AR/PP = K \times (U \times V_1/V_0)^m \qquad \ldots [4.15]$$

$$\frac{\text{Accumulated repair and maintenance costs, £}}{\text{Initial purchase price, £}} = \text{Repair constant}$$

$$\times \left[\text{Accumulated use, h} \times \frac{\text{Actual operating speed, km/h}}{\text{Typical operating speed, km/h}} \right]^{\text{(Repair exponent)}}$$

The values for the repair constant and for the repair exponent for different machines are given in Table 4.11. The indices of accumulated repair costs are based on the accumulated repair and maintenance costs to reduce the variability of the costs due to difference in the timing of the repairs. The *accumulated* use for tractors is given as engine hours divided by 1000, whilst that for machinery is given as operating hours divided by 1000. As the engine hours are recorded on a tractor hourmeter which is only correct at a particular engine speed for that tractor model, an under-utilised tractor, operating at low engine speeds, will record a lower hourmeter reading than a tractor operating at maximum power for the same period. Thus, the accumulated use based on the hourmeter reading partly accommodates for different levels of power utilisation in the calculation of tractor repair costs. The inclusion of an operating speed compensation ratio ensures that the fewer hours accumulated for a high speed operation do not result in lower repair costs than for a slower speed operation with the same machine. When the actual average operating speed is the same as the typical average operating speed, the speed ratio is equal to unity and has no effect on accumulated repair costs.

When *annual repair costs* are required on a year-by-year basis, they are calculated by subtracting the accumulated repair costs for the previous year, $AR_{(n-1)}$, from those for the current year, $AR_{(n)}$. So the annual repair costs, R_n, for the nth year, is given by the expression:

$$R_n = AR_{(n)} - AR_{(n-1)} \qquad \ldots [4.16]$$

Annual repair cost, £ = Accumulated repair cost for present year, £
  − Accumulated repair cost for previous year, £

For example, the accumulated repair costs from Table 4.12 for a grain drill are 127 per cent of the purchase price after 1500 h but only 54 per cent after 1000 h, so the additional repair costs for the last 500 h are 127 per cent less 54 per cent, namely 73 per cent of the purchase price.

Table 4.12   Repairs and maintenance costs as a percentage of purchase price for increasing accumulated use

Equipment	Repair and maintenance costs as a % of purchase price for increasing accumulated use, h						
	500	1000	1500	2000	4000	6000	8000
Tractor	0	1	2	5	20	38	82
Self-propelled combine harvester	3	12	28	51	—	—	—
Plough	12	43	89	150	—	—	—
Discs	6	18	36	58	—	—	—
Chisel plough	14	38	67	100	—	—	—
Grain drill	26	54	127	—	—	—	—
P.t.o. forage harvester	7	23	48	80	—	—	—

4.4.3 Labour charges

For 1986, the average cost – including National Insurance, Employer's Liability Insurance, overtime and benefits – is £117 a week for a general agricultural worker, £142 for a tractorman and £154 for a dairy stockman. Taking the average cost to the farmer for a tractorman, the hourly rate comes to £3.50 and the labour charge for machinery operations is simply this hourly rate multiplied by the time required.

This calculation for the labour charge infers that labour is available by the hour, on a casual basis, whereas the real value of labour at the time for which it is required may be considerably higher because of its opportunity cost. It may be necessary, for example, to hire permanent staff to ensure their availability at a critical time of the year. In such circumstances, the annual cost is more appropriate than the hourly labour cost. Provided that this aspect of labour availability is borne in mind, the hourly labour charge is an important component of machinery operating costs for systems comparisons at the enterprise level.

4.5 CALCULATION OF AVERAGE MACHINERY OPERATING COSTS

A quick, simple estimate of machinery operating costs is obtained by averaging the annual costs over the full period of ownership. This ignores the fact that depreciation is higher during the first year of ownership than in subsequent years, whilst repair and maintenance charges increase with age of the machine. Despite these approximations, the simplicity of the calculation is attractive and has ensured widespread adoption of the approach.

The basic economic data for estimating the average operating costs of two-wheel drive tractors are provided in Table 4.13. The average annual costs for different levels of annual use are given as a percentage of the purchase price, assuming that the tractors, in this example, are traded after a three-year period of initial ownership. With increasing annual use, the resale value is reduced. This affects both the level of depreciation and the interest charges. The major effect of the variation in annual use, however, is reflected in the repair charges which show a ten-fold increase between low use and high use over the three-year period. Actual values of the average annual costs are given in Table 4.14 for different sizes of two-wheel drive tractors used for

Table 4.13 The basic economic data for calculating the operational costs of two-wheel drive tractors at various levels of annual use, owned for 3 years

Economic data, % purchase price	Annual use, h/yr		
	500	1000	1500
Resale value after 3 years	66	57	51
Repairs after 3 years	2.73	11.16	25.42
Average annual costs			
Depreciation	11.33	14.33	16.33
Repairs	0.91	3.72	8.47
Interest at 4% net	3.32	3.14	3.02
Insurance at 1% av. investment	0.83	0.79	0.76
Shelter at 1% purchase price	1.00	1.00	1.00
Total	17.39	22.98	29.58

Fuel at £0.18 ℓ	0.55/kW h
Oil at £2.00/ℓ	$0.02169 + 0.00059 \times P_{max}$/h
Labour	£3.50/h
Capital cost	£2000 + 190 $\times P_{max}$
Power utilisation ratio	0.55

Table 4.14 The average annual costs and operating costs for different sizes of two-wheel drive tractors used for 1000 h/yr during the initial 3 years of ownership

	Average cost for different power units				
	30 kW	45 kW	60 kW	75 kW	90 kW
Annual costs, £/yr					
Depreciation	1011	1509	1916	2324	2731
Interest, insurance, shelter	380	520	661	801	942
Repairs	285	390	496	601	707
Fuel	1630	2450	3270	4080	4901
Oil	80	100	114	132	150
Total	3386	4969	6457	7938	9431
Tractor operating cost, £/h	3.39	4.97	6.46	7.94	9.43

1000 h/yr. Hourly operating costs, including labour, are shown in Fig. 4.10 for a range of annual usage. By spreading the fixed costs, the hourly operating costs decline with increasing annual use.

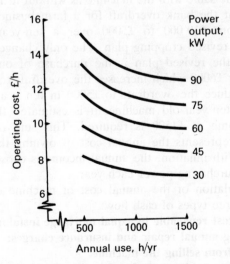

Fig. 4.10 Hourly operating costs, including labour, for different power sizes of two-wheel drive tractors.

This simple estimating procedure provides a useful guide to average trends but does not identify the variation in costs for different periods of ownership. The average operating costs are not strictly comparable with other annual farm costs and returns such as contractors' hire charges or crop gross margins. Neglecting to account for changes in costs from year to year also excludes the possibility of allowing for the effect of inflation. The correct evaluation of the annual machinery costs is particularly important to ascertain the economic life of a machine. For solutions to more complex machinery management problems, the annual machinery costs are calculated using the actual cash flows which occur each year.

4.6 ANNUAL MACHINERY COSTS FROM ACTUAL CASH FLOW

The profitability of an arable enterprise is usually determined on an annual basis by deducting the costs of producing a crop from the total returns. For this purpose, the annual cost of a machine is defined as the annual income which exactly balances the machine cost, so that, over its life, the change in the farm bank balance is the same with the machine as without it. For example, suppose that the bank overdraft for a farm business would be reduced from £10 000 to £5000 over a ten-year period by following a revised cropping plan. The only change required to implement the revised plan is the purchase of one additional machine for £6000 which increases the overdraft to £16 000. In order to reduce the overdraft to £5000 in ten years, including selling the ten-year-old machine, it is estimated that an extra annual income of £1500 is required. This sum of £1500, by definition, represents the annual cost of owning the additional machine. With inflation, the annual income is chosen so that it has equal purchasing power each year.

The calculation of the annual cost of machine ownership is based on three types of cash flow:
○ **capital cost repayable by equal mortgage instalments;**
○ **recurring annual repair and insurance charges;**
○ **income from selling the machine.**

The interest paid on the borrowed capital is included in the mortgage instalments but the effect of tax is considered separately because the different tax allowances and rates unnecessarily complicate the general calculation procedure.

Discounted cash flows are used to determine the current total cost in money of the same value. The essence of the discounting procedure is that present money is worth more than future money; the further ahead that the money is to be received in the future, the less it is worth in present-day terms. The reasons for this are two-fold: firstly, the further ahead that money is promised, the more risk that it may not materialise; secondly, cash received today can be invested to be worth more in the future. These reasons are valid even when the purchasing power of money remains constant but the effects of inflation can also be included in the calculation of actual cash flows.

Discounting has the opposite effect of compounding. A capital sun of £100 invested today at 8 per cent compound interest would be worth £216 at the end of ten years. Thus, £216 received in ten years' time, when discounted at 8 per cent, is worth £100 today or has a *present value* of £100. In discounted cash flow calculations, *discount factors* are used to translate the future value of money to its present value. The discount rate depends on either the overdraft rate of interest or on the opportunity cost placed on one's own capital allowing for risk. The present value of the net cash flow is then calculated by multiplying each annual net cash flow by the appropriate discount factor. The higher the rate of interest (or discount rate) and the further in the future that the money is to be received the lower the discount factor (Table 4.15). For example, the discount factor for year 10 at 8 per cent is 0.463 and, therefore, £216 in ten years time has a present value of £100 (£216 × 0.463 = £100).

Table 4.15 Discount factors for three rates of interest and three payback periods

Payback period, yr	Discount factors for interest rates		
	8%	12%	16%
1	0.926	0.893	0.862
5	0.681	0.567	0.476
10	0.463	0.322	0.227

To account for inflation, the net cash flows are increased by the expected rate of inflation and then discounted at the nominal rate of interest. Whilst this appears rather more laborious than discounting by the real rate of interest (i.e. the actual rate of interest adjusted for inflation) and produces the same end result, it nevertheless provides a more accurate assessment of the financial feasibility of the project, that is whether the cash returns are likely to be able to cover the repayment of capital and interest charges. Particularly when inflation and interest rates are high, the ability to service a loan in the early years may be difficult and the more complete procedure is advantageous.

4.6.1 Annual repayment of loan capital and interest

The capital cost of a machine, PP, may occur as a payment (outward cash flow) at the beginning of ownership at time zero, or else the machine may be bought by borrowing the money and paying a series of equal annual mortgage payments, M:

$$M = \frac{PP \times (i_l/100) \times (1 + i_l/100)^N}{(1 + i_l/100)^N - 1} \quad \ldots [4.17]$$

Annual mortgage payment, £ = Purchase price, £, compounded by Loan interest, %, for Period of ownership, yr

It is assumed that the period of the loan is the same as the period of ownership, N. Payment of one's own capital may be thought of as borrowing from onself at a lower interest rate. Thus, the concept of opportunity cost of capital may also be included.

The discounted cash flow or present cost of a cash flow, CF, in the year, n, is the amount of money, NPV, which must be invested now to pay for the cash flow in the nth year. If the interest on investments is i_i, then:

$$NPV = CF_n \times \frac{1}{(1 + i_i/100)^n} \quad \ldots [4.18]$$

Present cost of cash flow, £, from year n = Cash flow, £, in year n × Discount factor for year n

For a series of equal annual cash flows, M, over the life of the machine, the total present mortgage cost, NPV_m, is:

$$NPV_m = M \times \sum_{n=1}^{N} \frac{1}{(1 + i_i)^n} \qquad \ldots [4.19a]$$

Total present mortgage cost, £ = Annual mortgage payment, £ × Sum, Discount factors for each year of ownership

For the tabular method of calculating discounted cash flows (see section 4.6.5), eqn [4.19a] is used and the mortgage repayments can be obtained directly from Table 4.6. By rearranging terms and combining with eqn [4.17], the total present mortgage cost becomes:

$$NPV_m = PP \times \frac{i_l/100}{i_i/100} \times \frac{(1 + i_l/100)^N}{(1 + i_i/100)^N} \times \frac{(1 + i_i/100)^N - 1}{(1 + i_l/100)^N - 1}$$
$$\ldots [4.19b]$$

Total present mortgage cost, £ = Purchase price, £, compounded at loan interest rate and discounted at the investment interest rate for period of ownership

This is the investment needed at the start of ownership which pays for all the annual payments. If the interest rate on investments is the same as the loan interest rate, the total present mortgage cost is equivalent to the purchase price:

$$\text{If } i_i = i_l, \text{ then } NPV_m = PP \qquad \ldots [4.20]$$

Using own capital, Total present mortgage cost, £ = Purchase price, £

4.6.2 Recurring annual repair and insurance charges

The annual repair cost, R_n, for the nth year of machine ownership is obtained from eqn [4.16]. Insurance is charged at 1 per cent of the resale value of the machine at the end of previous year, $S_{(n-1)}$, from eqn [4.7].

If the rate of inflation is j, the present cost for repairs and insurance, NPV_r, is:

$$NPV_r = \sum_{n=1}^{N} [R_n + 0.01 \times S_{(n-1)}] \times (1 + j/100)^n \times \frac{1}{(1 + i_i/100)^n}$$
$$\ldots [4.21]$$

Total present cost of repairs and insurance, £ = Sum, Annual repair and insurance charges, £ × Inflation factor for year of ownership × Discount factor for year of ownership

4.6.3 Income from selling the machine

The current resale value of an N year old machine when the new machine is bought is S_N, so in N years time its value will have changed with inflation. The present resale value, NPV_s, after discounting is:

$$NPV_s = S_N \times (1 + j/100)^N \times \frac{1}{(1 + i_i/100)^N} \quad \ldots [4.22]$$

Present resale value, £ = Current resale value, £ × Inflation factor for period of ownership × Discount factor for period of ownership

4.6.4 Present annual ownership cost

The present annual ownership cost, A_o, is the value in today's money of N equal value, annual payments made during the ownership of the machine. These annual payments are again influenced by inflation and discounting, so that combining the three cash flows from eqn [4.18] plus eqns [4.20], [4.21] and [4.22]:

$$A_o \times \sum_{n=1}^{N} (1 + j/100)^n \times \frac{1}{(1 + i_i/100)^n}$$

$$= NPV_m + NPV_r - NPV_s \quad \ldots [4.23]$$

Sum, Present annual cost for each year of ownership, £ = Total present mortgage cost, £ + Total present repair and insurance cost, £ − Present resale value, £

If the inflated discount factor, w, is:

$$w = \frac{(1 + j/100)}{(1 + i_i/100)} \quad \ldots [4.24]$$

Inflated discount factor = Inflation factor × Discount factor

and since:

$$\sum_{n=1}^{N} w^n = w \times \frac{(w^N - 1)}{(w - 1)},$$

$$A_o = \left\{ NPV_m + \sum_{n=1}^{N} [R_n + 0.01 \times S_{(n-1)}] \times w^n - S_N \times w^N \right\}$$

$$\times \frac{(w-1)}{w \times (w^N - 1)} \qquad \dots [4.25]$$

Present annual ownership cost, £ = Total present mortgage cost, £ + Annual repair and insurance costs, £ − Resale value, £, adjusted for Inflation and Discount factors

4.6.5 Tabular method to calculate the annual cost of a machine

The tabular method is designed to calculate the annual ownership cost of a machine by completing a pro forma (Table 4.16). The following example illustrates the procedure.

Machine:	60 kW two-wheel drive tractor
Purchase price:	£13 400
Annual use:	1 000 h
Period of ownership:	5 years
Loan interest rate:	11%
Investment interest rate:	8%
Inflation rate:	5%

One line of the pro forma is completed for each year of ownership. The inflated cost of purchasing one of today's £ worth of goods is entered in column 2 of Table 4.16, by reference to Table 4.18. The amount to which £1 invested now will grow at the appropriate interest rate is entered in column 3, by reference to Table 4.18.

The annual repair costs are obtained from eqn [4.16], using the data from Table 4.11. The annual insurance costs are calculated by means of the resale values at the end of each previous year from eqn [4.7], using the factors listed in Table 4.4. These current annual costs for repairs and for insurance are listed in columns 4 and 5, respectively, and added together in column 6.

The current resale value at the end of the period of ownership is also obtained from eqn [4.7], using the factors listed in Table 4.4. This resale value is inserted in column 7, against the final year of use.

Table 4.16 Pro forma to calculate the annual cost of machine ownership, excluding tax relief

Col 1	Col 2	Col 3	Col 4	Col 5	Col 6	Col 7
Year	Inflation factor	Interest factor	Current repair cost, £	Current insurance cost, £	Current repair and ins cost, £	Current resale value, £
1	1.050	1.080	161	91	252	—
2	1.103	1.166	496	84	580	—
3	1.158	1.260	839	77	916	—
4	1.216	1.360	1186	71	1257	—
5	1.276	1.469	1537	65	1602	6006
6	—	—	—	—	—	—
7	—	—	—	—	—	—
8	—	—	—	—	—	—
9	—	—	—	—	—	—
10	—	—	—	—	—	—

Using Table 4.6, identify the annual mortgage repayment and enter this value, for each year of repayment, in column 8.

The actual repair and insurance costs in column 9 are obtained by multiplying the current repair and insurance costs in column 6 by the inflation factor in column 2. Similarly, the actual resale value in column 10 is obtained by multiplying the current resale value in column 7 by the inflation factor in column 2.

The actual cash outgoings, on an annual basis, are found from the sum of the annual mortgage payment and the actual repair and insurance costs less the actual resale value (column 8 + column 9 − column 10). These gross cash outgoings (excluding tax relief) are entered in column 11 and divided by the interest factor in column 3 to give the discounted cash outgoing in column 18. Columns 12 to 17 are discarded because they account for the tax allowances.

Col 8	Col 9 (Col 2 × Col 6)	Col 10 (Col 2 × Col 7)	Col 11 (Col 8 + Col 9 − Col 10)	Col 18 (Col 11 ÷ Col 3)	Col 19 (Col 2 ÷ Col 3)
Annual mortgage repayment, £	Actual repair and ins cost, £	Actual resale value, £	Gross cash outgoings, £	Discounted cash outgoings, £	Inflated discount factor
3626	265	—	3891	3603	0.972
3626	640	—	4266	3659	0.946
3626	1061	—	4687	3720	0.919
3626	1529	—	5155	3790	0.894
3626	2044	7664	−1944	−1357	0.869
—	—	—	—	—	—
—	—	—	—	—	—
—	—	—	—	—	—
—	—	—	—	—	—

				Box 18	Box 19
				13 415	4.600

Box 20

Annual cost of machine ownership, £   2 916

The sum of the discounted cash outgoings, in column 18, for every year of ownership, is listed in box 18. This calculates the right-hand side of eqn [4.23]. The inflated discount factors for each year are entered in column 19 by dividing the inflation factor by the interest factor (column 2 ÷ column 3). The sum of the inflated discount factors, in column 19, for every year of ownership is entered in box 19.

Dividing the sum of the discounted cash outgoings by the sum of the inflated discount factors (box 18 ÷ box 19) gives the annual cost of machinery ownership as shown in box 20.

4.6.6 Investment grants and taxation

Investment incentives and tax considerations affect farmers' decisions regarding the total investment in equipment and the

Table 4.17 Extended pro forma to calculate the annual cost of machine ownership, including tax relief (for use in conjunction with Table 4.16)

Col 1	Col 11	Col 12	Col 13 (Box 12 + Col 10 − PP)	Col 14	Col 15 (Col 9 + Col 12 + Col 14 − Col 13)	Col 16 (Col 14 × t)	Col 17 (Col 11 − Col 16)	Col 18 (Col 17 ÷ Col 3)	Col 19 (Col 2 ÷ Col 3)
Year	Gross cash outgoing, £	Actual capital allowance, £	Actual balancing charge, £	Actual interest charge, £	Total tax allowance, £	Total tax relief, £	Net cash outgoing, £	Discounted cash flow, £	Inflated discount factor
1	3891	3350	—	1474	5089	1527	2364	2189	0.972
2	4266	2513	—	1237	4390	1317	2949	2529	0.946
3	4687	1884	—	975	3920	1176	3511	2787	0.919
4	5155	1413	4484	683	3625	1088	4067	2990	0.894
5	−1994	1060	—	359	−1021	−306	−1688	−1149	0.869
6	—	—	—	—	—	—	—	—	—
7	—	—	—	—	—	—	—	—	—
8	—	—	—	—	—	—	—	—	—
9	—	—	—	—	—	—	—	—	—
10	—	—	—	—	—	—	—	—	—
		Box 12 10 220					Box 18 9 346	Box20 2 032	Box19 4.600

Annual cost of machine ownership, £

Table 4.18 Capital growth per £100

Term, year	Capital growth per £100 for various rates of interest/inflation, %														
	4	5	6	7	8	9	10	11	12	13	14	15	16	18	20
1	104.00	105.00	106.00	107.00	108.00	109.00	110.00	111.00	112.00	113.00	114.00	115.00	116.00	118.00	120.00
2	108.16	110.25	112.36	114.49	116.64	118.81	121.00	123.21	125.44	127.69	129.96	132.25	134.56	139.24	144.00
3	112.48	115.76	119.10	122.50	125.97	129.50	133.10	136.76	140.49	144.28	148.15	152.08	156.08	164.30	172.80
4	116.98	121.55	126.24	131.07	136.04	141.15	146.41	151.80	157.35	163.04	168.89	174.90	181.06	193.87	207.36
5	121.66	127.62	133.82	140.25	146.93	153.86	161.05	168.50	176.23	184.24	192.54	201.13	210.03	228.77	248.83
6	126.53	134.00	141.85	150.07	158.68	167.71	177.15	187.04	197.38	208.19	219.49	231.30	243.63	269.95	298.59
7	131.59	140.71	150.36	160.58	171.38	182.80	194.87	207.61	221.06	235.26	250.22	266.00	282.62	318.54	358.31
8	136.85	147.74	159.38	171.81	185.09	199.25	214.35	230.45	247.59	265.84	285.25	305.90	327.84	375.88	429.98
9	142.33	155.13	168.94	183.84	199.90	217.18	235.79	255.80	277.30	300.40	325.19	351.78	380.29	443.54	515.97
10	148.02	162.88	179.08	196.71	215.89	236.73	259.37	283.94	310.58	339.45	370.72	404.55	411.14	523.38	619.17
11	153.94	171.03	189.82	210.48	233.16	258.04	285.31	315.17	347.85	383.58	422.62	465.23	511.72	617.59	743.00
12	160.10	179.58	201.21	225.21	251.81	281.26	313.84	349.84	389.59	433.45	481.79	535.02	593.60	728.75	891.61
13	166.50	188.56	213.29	240.98	271.96	306.58	345.22	388.32	436.34	489.80	549.24	615.27	688.57	859.93	1069.93
14	173.16	197.99	226.09	257.85	293.71	334.17	379.74	431.04	488.71	553.47	626.13	707.57	798.75	1014.72	1283.91
15	180.09	207.89	239.65	275.90	317.21	364.24	417.72	478.45	547.35	625.42	713.79	813.70	926.55	1197.37	1540.70
20	219.11	265.32	320.71	295.21	466.09	560.44	672.74	806.23	964.62	1152.30	1374.34	1636.65	1946.07	2739.30	3833.75

timing of individual machinery purchases. *Investment grants* are made available to farmers as an incentive for further mechanisation. These grants are a straightforward payment which reduces the net cost of a machine and can be considered as a Government discount to the buyer. This form of Governmental incentive benefits all farmers equally, irrespective of the level of business profitability or taxable income. These grants are being decreased or discontinued with the shift in emphasis from increasingly efficient production techniques to conservation of the rural environment.

Capital investment in buildings and machinery is also eligible for tax relief, by means of an annual *capital allowance*. In contrast with the investment grant, capital allowances only benefit farmers who make sufficient profit to pay tax. The more profitable the business, the higher the marginal tax rate and the greater the benefit from the capital allowances. In discounted cash flow terms, the tax relief on annual capital allowances is worth progressively less relative to an investment grant of the same total amount received soon after the time of purchase. The effect of various tax allowances can be more readily understood by extending the example using discounted cash flows (section 4.6.5) to calculate the present annual ownership costs for a machine taking into account tax relief at the standard rate.

For taxation purposes, the annual rate of capital allowance for a building is 4 per cent of the purchase price, using straight line depreciation over 25 years. For machinery, the annual rate of capital allowance is 25 per cent on a diminishing balance basis, that is on the written down value of the machine. Thus, the annual capital allowance for an 'n' year old machine, CA_n, is

$$CA_n = (25/100) \times (1 - 25/100)^{(n-1)} \times PP \qquad \ldots [4.26]$$

Annual capital allowance, £ = Annual rate of capital allowance, %
× Diminishing balance of Purchase price, £

Once the machine is eventually sold, or traded in, the total capital allowance must be adjusted to equate with the actual loss in value of the machine during the period of ownership. If the resale value exceeds the written-down value, then it is necessary to have a *balancing charge* on which tax must be paid. This balancing charge in the last year of ownership, BC_N, is:

$$BC_N = \sum_{n=1}^{N} CA_n + S_N - PP \qquad \dots [4.27]$$

Balancing charge, £ = Sum of Annual capital allowances, £ + Resale value, £
— Purchase price, £

Alternatively, if the resale value is less than the written-down value, then there is a *balancing allowance* (i.e. a negative balancing charge) on which additional tax relief is available. In practice, the balancing charge or allowance is deducted from or added to the capital allowances on other machines within the farm equipment 'pool'.

Other machinery costs eligible for tax relief are:
interest payments;
repairs and insurance charges;
fuel and oil.

When repaying the purchase price of a machine by means of a mortgage in equal instalments, the initial instalments largely comprise interest whilst later instalments are mainly repayment of the principal. The annual interest charge, I_n, is given by the interest on the outstanding balance of the loan after repayment of the mortgage instalments in the preceding period of the loan:

$$I_n = \frac{PP \times (1 + i_l/100)^N - (1 + i_l/100)^{(n-1)}}{(1 + i_l/100)^N - 1} \times (i_l/100) \quad \dots [4.28]$$

Annual interest charge, £ = Oustanding balance, £ × Loan interest rate, %

Repairs and insurance costs have already been determined previously (eqn [4.21]) and fuel costs can be considered separately as they are already in present value terms.

The various tax allowances are multiplied by the marginal tax rate, t, to give the tax relief. There is a series of taxable income bands, each with its own tax rate, ranging from the standard tax rate of 30 per cent up to 60 per cent at higher levels of taxable income. The annual tax relief is deducted from the gross cash outgoings to give the net mounts for discounting. Extending eqn [4.25] to include tax relief, the present annual ownership cost is given by the relationship:

$$A_t = \{NPV_m - (t/100) \times \sum_{n=1}^{N} (CA_n + I_n)/(1 + i_i/100)^n$$

$$+ (t/100) \times BC_N/(1 + i_i/100)^N$$

$$+ (1 - t/100) \times \sum_{n=1}^{N} [R^N + 0.01 \times S_{(n-1)}] \times w^n$$

$$- S_N \times w^N\} \times (w - 1)/[w \times (w^N - 1)] \quad \ldots [4.29]$$

Present annual ownership cost after tax, £ = Total present mortgage cost, £ − Tax relief on annual Capital allowances and Interest, £ + Tax repaid on Balancing charge, £ + Annual repairs and insurance costs after tax relief, £ − Resale value, £ adjusted for Inflation and Discount factors

The pro forma in Table 4.16 is extended in Table 4.17 to include tax allowances. The gross cash outgoings from Table 4.16 are repeated in Table 4.17. The capital allowances for each year of ownership are calculated from eqn [4.26] and entered in column 12. The sum of these capital allowances in column 12 for the whole period of ownership is entered in box 12 and used in eqn [4.27] to calculate the balancing charge for the final year of ownership which is entered in the last line of column 13. Annual interest charges from eqn [4.28] are inserted in column 14.

All the tax allowances for repairs and insurance, for capital investment, for interest and less the balancing charge (column 9 + column 12 + column 14 − column 13) are entered in column 15. These values multiplied by the marginal tax rate give the total tax relief in column 16. In this example, the standard tax rate of 30 per cent is used. The actual cash outgoings in column 17 are the gross cash outgoings less the tax relief for each year (column 11 − column 16).

As in the previous example, the actual cash outgoings are individually discounted in column 18. The sum of the discounted cash flows in box 19 is divided by the sum of the inflated discount factors in box 20 to give the present annual ownership costs for the machine after tax.

The present annual cost of machine ownership is substantially altered by tax considerations. In the example, the effect of allowing for tax reduces the present annual cost of tractor ownership from £2916 to £2032, a reduction of 30 per cent which, by coincidence, is the same value as the tax rate used. For the same tractor, and considering the same fixed and running costs, the average annual ownership cost is £2440 using the simple procedure outlined in section 4.5. This latter approach yields an intermediate cost figure within the range spanned by the present annual ownership costs with and without tax. There is, however, both substantial room for error and an absence of information on cash flows compared with the discounted cash flow approach which has the further flexibility of incorporating tax allowances.

<h2 style="text-align:center">4.7 SUMMARY</h2>

Fixed costs, mainly dependent on duration of ownership, are: depreciation; interest; tax and insurance; shelter.

Running costs, proportional to utilisation, are: fuel and oil; repairs and maintenance; labour.

Machinery purchase prices, in many cases, are linearly related to size through major design features such as engine power rating for tractors, number of furrows for ploughs, etc. The purchase prices can be index linked to inflation.

The resale value of a machine is determined by the effect of age and level of use on the rate of decremental depreciation.

Net interest is a simple concept which minimises the effect of inflation when calculating average operating costs for machinery.

Vehicle excise duty and insurance charges are 1 per cent of the market value of a machine.

Shelter charges are taken as 1 per cent of the purchase price of a machine.

Repair and maintenance charges, as a proportion of the purchase price of a machine, increase logarithmically with the accumulated use of the equipment during the period of ownership.

Fuel consumption for tractors assumes an average power utilisation of 55 per cent of maximum power-take-off power throughout the period of operation.

Average operating costs for machines are readily calculated but tend to underestimate the high level of ownership costs in early years.

Calculation procedures for the present annual cost of machine ownership from actual cash flows provide a detailed appraisal of both the annual cash flow throughout the period of ownership and the opportunity to include tax allowances.

5

INDIRECT MACHINERY COSTS

OUTLINE

Machinery sheduling; Yield losses through untimely establishment, establishment timespan and optimum date, planting loss chart; Timing of spray applications; Grain losses through untimely harvesting, shedding and cutter-bar losses, gale damage, spread of ripening, separation loss; Valuation of forage losses, livestock output forgone; Soil weathering benefits.

APOSTROPHE – PLOUGH MONDAY A MEDIAEVAL TRICK OR TREAT

Plough Monday, the first Monday after the Twelfth Day of Christmas, was once a day of great festivity all over England. The chief feature of festivals was the Fool Plough decorated with ribbons – and there was considerable competition among ploughmen who wanted to be chosen as 'plough bullocks', so that they could pull the plough in a procession that wended its way round each locality. The object of the procession was to collect money and the leader of the throng, a man dressed as an old woman known as 'Bessy', carried the collecting box. Behind him came the plough drawn by 30 or 40 gaily decorated ploughmen and other followers blowing cow horns or clattering the tools of their trade. There would be threshers carrying flails, reapers bearing sickles, carters with long whips, blacksmiths and millers.

Even the poorest cottagers managed to contribute a coin but strategic halts were made at large farm houses, where it was customary for the principals to be rewarded with food and ale. A negative response from a house of obvious well-being led to more extreme tactics being employed. The procession would swarm into the grounds with much yelling and blowing of cow horns. Further failure to respond was often followed by a demonstration of the ploughmen's prowess and it did not take long for carefully tended grounds to resemble a newly-ploughed field. Such extreme action seldom had to be repeated for the ploughmen's charter was an ancient right beyond the law and there is no evidence that this right was ever successfully challenged by an unfortunate victim. The festivities ended when the principals had supped their 'load' of ale and the popularity of the day is credited to the need for a little fun to break the monotony of a dreary winter scene.

(Source: Humphreys)

5.1 INTRODUCTION

An *indirect machinery cost* is the financial loss which is incurred through inadequate scheduling of machinery operations, causing a reduction in crop yield or quality. The most important indirect machinery charges are crop losses through:

○ **untimely establishment;**
○ **untimely spraying;**
○ **untimely harvesting;**
○ **excessive soil compaction.**

Just occasionally, there may be an *indirect machinery saving* accrued through a reduction in crop production costs, for example, by premature scheduling of machinery operations to accommodate:

○ **timely soil weathering**

which reduces cultivation requirements.

Timely operations in crop production have a major impact on the economic viability of both arable and livestock enterprises. Late establishment of an arable crop curtails the growing season and reduces the potential yield or introduces a greater risk from harvesting in more unfavourable conditions towards the end of the year. Delays in crop spraying may reduce the effectiveness of a herbicide treatment on more mature plants or may dramatically alter the population dynamics of an insect attack or disease infestation. Late harvesting of potatoes leads to a greater incidence of storage diseases which reduce saleable yield. Postponing the start of mowing produces a greater bulk of forage but at the expense of digestibility, so that animal liveweight gain is impaired.

In consequence, the vital importance of adequate machine capacity is universally recognised but the required adequacy is difficult to assess. To compensate for the vagaries of the weather, the commonest approach to machine capacity is investing in a healthy reserve. Whilst the extra machine operating costs are acknowledged, scant consideration is given to the indirect machinery costs. The over-investment in a machine could lead to finishing work before the optimum date of completion – though this seldom happens because of our propensity to procrastinate when there is time in hand! Early establishment of spring crops, for example, may increase the risk of seed damage in a cold, wet environment and decrease the potential yield to as great an extent as late establishment. Mistimed chemical treatment, when a crop is not in jeopardy, is equivalent to a routine programme of prophylactic spraying which, though commonly practised to minimise the level of risk in crop production, is being challenged by conservation groups as condoning chemical abuse. More frequent cutting of grass seldom increases forage quality sufficiently to compensate financially for the reduction in quality except perhaps for specialised grass drying operations.

In addition to those penalties from untimely operations, higher investment in machines is almost always translated into heavier units which have the potential to cause much greater soil damage. Whilst crop yield responses to traffic density varies with soil properties and plant species, the simplest valuation of a compaction penalty is the cost and frequency of remedial subsoiling work.

Only in one case does there appear to be any real opportunity to achieve an indirect machinery saving. For spring-sown crops, the completion of primary tillage before the onset of winter provides time for soil weathering or frost moulding which may reduce the operations required for seedbed preparation.

The indirect machinery charges for cash crops are easy to assess from the loss in price of saleable crops. Forage crops, however, are not sold and their loss in feeding value must be translated into costs of livestock output forgone which is more subjective because it involves ration formulation and feed conversion ratios to determine the financial implications.

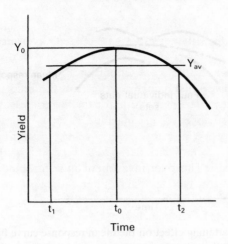

Fig. 5.1 General form of the yield response curve, showing the peak yield and the average yield over a timespan.

5.2 CROP YIELD LOSSES FROM UNTIMELY ESTABLISHMENT

The results of numerous individual experiments confirm that crop yield varies in a predictable way in relation to the timing of various field operations (Fig. 5.1). The combination of geographical location, soil type, seasonal weather variation and crop variety ensures that the crop response from each experiment is unique. The general pattern which emerges, however, is that the yield response is curvilinear. Thus, the optimum date on which an operation should be performed is when the crop yield function reaches a maximum value. Unless the enterprise is on a very small scale or the machine used is very large, it is unlikely that the operation can be completed at that specific time. This inability to complete an operation within a short period incurs a penalty which increases on a daily basis as the duration of the period is extended. The evaluation of these penalty costs requires the selection of a unique yield/time response for a multiplicity of crop yield experiments.

5.2.1 Timeliness coefficients

Traditionally, the results of individual crop yield experiments for

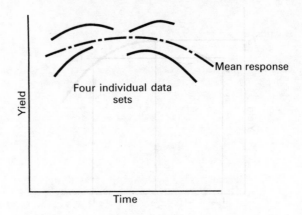

Fig. 5.2 The flattening effect on the mean response curve by combining data sets from different experiments.

different years and/or sites are combined into a mean yield curve by applying regression analysis (Fig. 5.2). This mean curve is much flatter with respect to time than the curves for any individual experiment. The value of the maximum yield has a high statistical significance by this analysis but the optimum date of establishment is very suspect because of the flatness of the curve.

Instead of plotting the Julian calendar date on the abscissa, time may be measured from any significant event in plant development such as planting, emergence and so on. In different seasons the same significant event in plant development can occur over a range of calendar dates. These seasonality effects are decreased by normalising the experimental yield results about the date of crop establishment for peak yield. Some of the varietal and site differences are also decreased by transforming the experimental yield data into percentage yield loss relative to the peak value for each trial. These percentage yield losses are predicted by two independent equations, one for early losses and one for late losses, each having the general form:

$$y = K_t \, (t_o - t)^2 \qquad \qquad \ldots \text{[5.1]}$$

Yield loss, % $\propto$ Timeliness coefficient $\times$ Time deviation from optimum, Julian day number

As the two timeliness coefficients, K_{t1} for early operations and K_{t2} for late operations, can have different values, the percentage

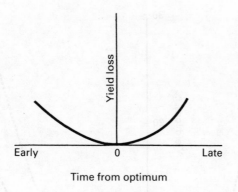

Fig. 5.3 The general form of the yield loss curve, normalised for time from the optimum date of establishment.

yield loss curves are not necessarily symmetrical on both sides of the optimum time (Fig. 5.3). At the optimum date for crop establishment, there is a smooth transition from one curve to the other because the gradients of both curves are zero at that point. Establishment dates are all converted to Julian day number, day 1 being 1 January. The timeliness coefficients from an analysis of almost 3500 experimental yield measurements for eight separate crops are listed in Table 5.1.

When the two timeliness coefficients have a similar value for a crop, as in the case of winter wheat and swedes, the composite curves are virtually symmetrical about the ordinate (Fig. 5.4).

Table 5.1 Timelines coefficients for early and late establishment of eight crops

Crop	Timelines coefficient $\times 10^{-3}$	
	Early	Late
Winter barley	3.10	3.84
Winter wheat	4.44	4.35
Spring barley	9.11	11.0
Spring wheat	8.78	10.9
Oats	13.5	19.4
Potatoes (maincrop)	5.81	9.13
Swedes	17.2	18.4
Turnips	49.6	31.7

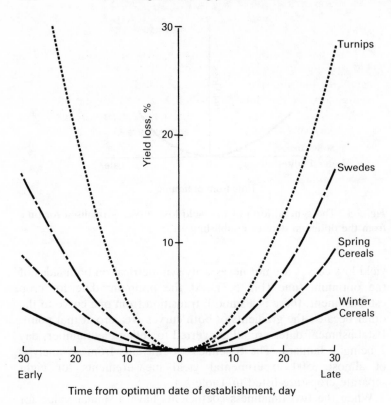

Fig. 5.4 Percentage yield losses from untimely crop establishment.

The losses for these crops are, therefore, similar for both early and late establishment. The turnip crop suffers a substantially greater percentage yield loss through early establishment. The yields of all other crops (winter barley, spring barley, spring wheat, oats and potatoes) suffer slightly greater reductions with late establishment.

Winter cereals have percentage yield loss curves which are much flatter than those for spring cereals and potatoes, whilst the yields of fodder root crops are most sensitive to time of establishment. This is borne out in practice by farmers switching from turnips to swedes if a delay in sowing becomes unavoidable.

5.2.2 Overall yield losses

The overall yield loss is obtained by integrating the percentage yield loss equations (eqn [5.1]) with the relevant timeliness coefficients for the timespan. The level of the percentage yield losses is influenced by the timespan which is chosen for the machinery operations and there are three optional management schedules:

○ **premature scheduling;**
○ **late scheduling;**
○ **balanced scheduling.**

When the machinery operation is planned to be completed before the optimum time, the policy is called *premature scheduling*, and might be used with cultivation work in order to achieve an optimum date for cereal drilling. For premature scheduling when the work starts at date t_1 and finishes at date t_2, the overall yield loss as a percentage is:

$$y_{ov} = K_{t1} \times (t_2 - t_1)^{2/3} \qquad \ldots [5.2a]$$

Overall yield loss, % ∝ Early timeliness coefficient × Timespan, days

Similarly, *late scheduling* is appropriate for soft fruit harvesting which cannot commence until the crop reaches maturity but then proceeds as quickly as possible to avoid deterioration of a perishable commodity. The overall yield loss for late scheduling as a percentage is:

$$y_{ov} = K_{t2} \times (t_2 - t_1)^{2/3} \qquad \ldots [5.2b]$$

Overall yield loss % ∝ Late timeliness coefficient × Timespan, days

For crop establishment, the lowest percentage yield loss is obtained through *balanced scheduling*. In this case, work starts early at date t_1 before the optimum date, t_0, and finishes late at date t_2 after the optimum date (Fig. 5.1). The overall yield loss then becomes the means of both the early and late percentage yield losses:

$$y_{ov} = \tfrac{1}{3}[K_{t1} \times (t_0 - t_1)^3 + K_{t2} \times (t_2 - t_0)^3]/(t_2 - t_1) \quad \ldots [5.2c]$$

Overall yield loss, % ∝ Early yield loss, % + Late yield loss, %

The overall yield loss for balanced scheduling is, therefore, one third of the marginal yield losses at the start of the establishment period and at the finish.

Table 5.2 The duration of the establishment period before and after peak yield for various average levels of the percentage yield loss for arable crops throughout the UK

Yield loss, %		Duration of establishment					
		Winter barley		Winter wheat		Spring barley	
Overall	Margin	Early	Late	Early	Late	Early	Late
0.5	1.5	22	20	18	19	13	12
1.0	3.0	31	28	26	26	18	16
1.5	4.5	38	34	32	32	22	20
2.0	6.0	44	40	37	37	26	23
2.5	7.5	49	44	41	42	29	26
3.0	9.0	54	48	45	45	31	29

5.2.3 Time penalty costs

For a given overall yield loss in the arable areas throughout the United Kingdom, the duration of the establishment period before and after the date of peak yield is presented in Table 5.2 for various crops; the greater the overall yield loss which is considered acceptable, the longer the timespan for establishing the crop and the smaller the capacity of the equipment required. The time penalty cost, *TC*, due to the untimely establishment of a crop is given by the value of this lost yield:

$$TC = Y_o \times y_{ov} \times A \times p_c \qquad \ldots [5.3]$$

Time penalty cost, £ = Peak yield, t/ha × Overall yield loss, %
× Crop area, ha × Crop price, £/t

The expected peak crop yield and the optimum date of crop establishment are often known for a particular location. Alternatively, average values which were obtained by further analysis of the experimental dates are presented in Table 5.3. Using either the local data or average results, the time penalty costs can be added to the machinery operating costs to select the optimum size of machinery required for crop establishment.

5.2.4 Relating yield loss to machinery rate of work

The yield loss from delayed planting of maincrop potatoes may

period before and after peak yield for various crops, days									
Spring wheat		Oats		Potatoes		Swedes		Turnips	
Early	Late	Early	Late	Early	Late	Early	Late	Early	Late
13	12	11	9	16	13	9	9	5	7
18	17	15	12	23	18	13	13	8	10
23	20	18	15	28	22	16	16	10	12
26	23	21	18	32	26	18	18	11	14
29	26	23	20	36	29	21	20	12	15
32	29	26	22	39	31	23	22	13	17

be related to the work rate of the potato planter by means of a chart. For this application, it is assumed that there is no yield loss from early planting up to a critical date, after which a linear yield loss is incurred. The critical date of 13 April on the chart is almost identical with the date of optimum yield in Table 5.3. Although the daily yield loss of 0.25 t/ha is twice that indicated in Table 5.2, the approach remains relevant.

The yield loss sustained through failing to plant the whole area by the critical date (13 April) will depend on the area remaining to be planted and the average area planted daily, which jointly will govern the number of days required to finish planting. The average area planted daily is based on the seasonal rate of work, taking into account weekends *and* wet weather. Usually, the calendar rate of work is rather more than half the overall rate of work. Thus, for four days of actual planting spread over seven days, the calendar rate of work is 4/7 of the overall rate of work.

As an example of the yield loss calculation, assume that 14 ha of potatoes remain to be planted on the evening of 13 April. If the calendar rate of planting is 2 ha/day, a further seven days will be required to complete the job which will finish on 20 April. The cumulative loss in yield, sustained over the 14 ha, is the sum of the daily losses. At the end of the first day after the critical date, 2 ha will be planted and the yield loss for that area will be:

$$0.25 \text{ t ha}^{-1} \text{ day}^{-1} \times 2 \text{ ha} \times 1 \text{ day} = 0.5 \text{ t}$$

Table 5.3 The optimum establishment date, day number and peak yield for eight crops at various locations

	Winter barley	Winter wheat	Spring barley	Spring wheat	Oats	Potatoes	Swedes	Turnips
Locations: all sites								
Optimum sowing date	15 Oct	22 Oct	18 Mar	18 Mar	23 Mar	14 Apr	5 May	18 May
Optimum day number	288	296	76	76	81	104	125	138
Peak yield (t/ha)	5.95	6.20	4.88	3.83	4.92	42.21	5.27*	6.44*
Location: Bush, Edinburgh								
Optimum sowing date	4 Oct	1 Oct	28 Mar	NA	NA	NA	28 Apr	NA
Optimum day number	77	274	86				118	
Peak yield (t/ha)	6.18	8.87	6.03				5.52*	
Location: Craibstone, Aberdeen								
Optimum sowing date	30 Sept	NA	16 Mar	NA	23 Mar	12 Apr	5 May	18 May
Optimum day number	273		74		81	102	125	138
Peak yield (t/ha)	6.48		5.06		4.92	42.48	4.00*	6.44*
Location: Arthur Rickwood, Cambridge								
Optimum sowing date	7 Nov	28 Oct	NA	15 Mar	NA	NA	NA	NA
Optimum day number	311	301		73				
Peak yield (t/ha)	5.03	5.33		4.07				
Location: Boxworth, Cambridge								
Optimum sowing date	21 Oct	29 Oct	28 Mar	23 Mar	NA	NA	NA	NA
Optimum day number	294	302	86	81				
Peak yield (t/ha)	6.16	5.93	4.37	3.67				

* Yield of dry matter, t/ha
NA Not available

On the second day, a further 2 ha will be planted with a yield loss of:

$$0.25 \text{ t ha}^{-1} \text{ day}^{-1} \times 2 \text{ ha} \times 2 \text{ day} = 1 \text{ t}$$

and so on until the seventh day when the losses for the last 2 ha will be:

$$0.25 \text{ t ha}^{-} \text{ day}^{-1} \times 2 \text{ ha} \times 7 \text{ day} = 3.5 \text{ t}$$

The sum of all the daily losses will be:

$$3.5 + 3.0 + 2.5 + 2.0 + 1.5 + 1.0 + 0.5 = 14 \text{ t}$$

As the area planted each day is equivalent to the calendar rate of work, the cumulative loss of crop, Y_T, can be reckoned more simply from the total delay period, X_1, from the equation:

$$Y_T = y_d \times C_c \times (X_1^2 + X_1)/2 \qquad \ldots [5.4]$$

Cumulative yield loss, t $\propto$ Daily yield loss, t ha^{-1} day^{-1} $\times$ Calendar rate of work, ha/day $\times$ Delay period, day

The results of such calculations are given in Table 5.4 for a range of calendar rates of work for potato planting over a considerable number of delay periods. Calendar rates of work for planting in excess of 5 ha/day are appropriate for large farms or producer groups where there may be more than one planter in use. For calendar rates of work below 1 ha/day, the yield losses can be obtained by moving the decimal point in the appropriate column of Table 5.4.

The data from this table are expressed graphically in Fig. 5.5 which facilitates the comparison between planters working at different rates. In the chart, it is assumed that there are 16 calendar days for potato planting from 29 March, the earliest date (on average) with low risk of frost damage. The calendar rate of planting is shown by the lines which radiate from the common starting point. The horizontal lines across the chart show the crop area to be planted. The date on which planting would be completed, if started on 29 March, is directly below the point where the 'workrate line' cuts the 'area line'. The thick vertical line, A–A, marks the night of 13 April, and later completion dates (to right of this line) fall in the area where the loss is read off the contour 'loss line', centring on the top right-hand corner of the chart.

Table 5.4 The cumulative loss of potato crop yield likely to result from a delay in planting at different calendar rates of work and based on a daily yield loss of 0.25 t/ha (adapted from ADAS, 1976)

Date	Time to complete planting, day	Cumulative yield loss, t, at different calendar work rates for planting, ha/day													
		1.0	1.5	2.0	2.5	3.0	3.5	4.0	4.5	5	6	7	8	9	10
April															
14	1	0.25	0.375	0.50	0.65	0.75	0.875	1	1.13	1.25	1.50	1.75	2	2.25	2.50
15	2	0.75	1.125	1.50	1.88	2.25	2.63	3	3.38	3.75	4.50	5.25	6	6.75	7.50
16	3	1.50	2.25	3.00	3.75	4.50	5.25	6	6.75	7.50	9.00	10.50	12	13.50	15.00
17	4	2.50	3.75	5.00	6.50	7.50	9.75	10	11.25	12.50	15.00	17.50	20	22.50	25.00
18	5	3.75	5.60	7.50	9.40	11.30	13.10	15	16.90	18.80	22.50	26.25	30	33.75	37.50
19	6	5.25	7.90	10.50	13.10	15.75	18.40	21	23.60	26.25	31.50	36.75	42	47.25	52.50
20	7	7.00	10.50	14.00	17.50	21.00	24.50	28	31.50	35	42	49	56	63	70
21	8	9.00	13.50	18.00	22.50	27.00	31.50	36	40.50	45	54	63	72	81	90
22	9	11.25	16.90	22.50	28.10	33.80	39.40	45	50.60	56	68	79	90	101	113
23	10	13.75	20.60	27.50	34.40	41.30	48.10	55	61.90	69	83	96	110	124	138
24	11	16.50	24.80	33	41.30	49.50	57.80	66	74	83	99	115	132	148	165
25	12	19.50	29	39	49	59	68	78	88	97	117	136	156	175	195
26	13	22.75	34	45	57	68	80	91	102	114	136	159	182	205	228
27	14	26.75	39	52	66	79	92	105	118	131	157	184	210	236	262
28	15	30.00	45	60	75	90	105	120	135	150	180	210	240	270	300
29	16	34.00	51	68	85	102	119	136	153	170	204	238	272	306	340
30	17	38.25	57	76	96	115	134	153	172	191	229	268	306	344	382

May														
1	18	42.75	64	85	107	128	150	171	192	214	256	299	342	385	427
2	19	47.50	71	95	119	142	166	190	214	237	285	332	380	427	475
3	20	52.50	79	105	131	157	184	210	236	262	315	367	420	472	525
4	21	57.75	87	115	144	173	202	231	260	289	346	404	462	520	577
5	22	63.25	95	126	158	190	221	253	285	316	379	443	506	569	632
6	23	69	103	138	172	207	241	276	310	345	414	483	552	621	690
7	24	75	113	150	188	225	263	300	338	375	450	525	600	675	750
8	25	81	122	162	203	244	284	325	366	406	487	569	650	731	—
9	26	88	132	175	219	263	307	351	395	439	526	614	702	—	—
10	27	95	142	189	236	283	331	378	425	472	567	661	—	—	—
11	28	102	152	203	254	304	355	406	457	507	609	—	—	—	—
12	29	109	163	217	272	327	381	435	489	544	—	—	—	—	—
13	30	116	174	232	290	348	406	465	522	—	—	—	—	—	—

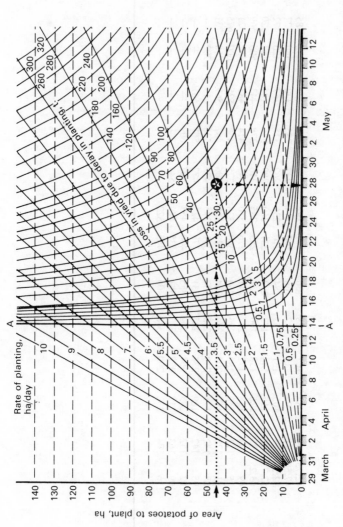

Fig. 5.5 The likely loss of production through low work rate delaying the completion of planting beyond the latest date, here taken as 14 April (line A–A), for maximum yield (adapted from ADAS, 1976). For example: the star identifies the point where the horizontal 'area to plant' line (45 ha) cuts the radial 'planting rate' line (1.5 ha/day); the estimated completion date for planting is vertically below the star (28 April, a delay of 14 days) and the 'loss' curve closest to the star indicates the probable production loss (40 t).

The chart aids machinery management by showing the effect of various planting rates on the date of completion of planting and on the yield loss due to late planting. Thus, if 40 ha of potatoes are to be planted at a calendar rate of work of 2 ha/day, planting will be completed on 18 April, by which date a loss of yield of 5 t will have occurred. Had the same 40 ha been planted by a machine with a calendar rate of work of 3 ha/day, the operation would have been completed by 12 April, and there would have been no loss of yield. Under those circumstances, the second machine, through its faster speed, would have generated extra revenue from five tonnes of potatoes which, at £60/t, is £300 or £7.50/ha over the whole area.

The revenue generating potential, of course, varies with the area of the crop. For a smaller area of only 25 ha of potatoes, both planters would have completed the work before the critical date of 14 April (on 10 and 6 April, respectively) and neither would have caused any loss through late planting. By contrast, 66 ha would require 22 days work with the faster machine, finishing on 19 April with a yield loss of 15 tonnes; work with the smaller machine must continue until 30 April, by which date the loss would become 75 tonnes. The difference between the alternatives is now 60 t over the whole 66 ha, worth £3600 or £55/ha.

The chart may be adapted, quite readily, to meet particular needs. When the start of planting in an area of heavy land is commonly expected to be delayed by a week until 5 April, the whole array of radiating lines on the chart can be moved seven days to the right relative to the rest of the chart. The later starting date would result in a greater loss of yield, unless a faster machine was obtained. A further possibility is where the yield penalty commences at a later critical date, such as for more northerly or higher altitude locations. In this case, the calendar dates are altered along the baseline but the format of the chart remains the same.

5.3 INFLUENCE OF TIMING OF FUNGICIDAL SPRAYS ON CEREAL YIELDS

The most important leaf disease of barley is powdery mildew which, if uncontrolled, would have considerable economic impact over the 2 million hectares of the crop grown annually in the UK. The effect of mildew on crop yield can be avoided or reduced by the use of either fungicides or more resistant crop varieties. For both economic and conservationist reasons, it is desirable to minimise the number of spray applications required. Single sprays of fungicides can give satisfactory control of mildew and increase crop yields but the timing for a single spray is more critical than for multiple applications.

In a series of 56 trials during 1971–73, single fungicide sprays of tridemorph (0.52 a i/ha) were applied at rates of 220–440 l/ha, at different times, to five different varieties of spring barley, the most common being Zephyr which is known to be susceptible to mildew. Figure 5.6 illustrates the increases from single sprays applied at the first record of mildew, at the first record of 3, 5 and 10 per cent mildew on leaf 3 or leaf 4 and also the yield increases from single sprays applied up to four weeks before and after those application times. These data indicate that crop yield increases are obtained from sprays applied over a fairly long period of time in relation to any one particular level of mildew. The best response, however, is when sprays are applied as soon as 3–5 per cent mildew is recorded. This optimum timing was irrespective of the growth stage of the crop or the proximity of winter barley and the consequential 'green bridge' effect which encourages the more rapid development of the disease through carryover of infection on growing crops from one season to the next. Sprays applied one to four weeks before or after the 3–5 per cent mildew stage were less effective. Spraying at the first record of mildew is a week too early (Fig. 5.6(a)) and spraying with over 10 per cent mildew present is a week too late (Fig. 5.6(d)).

An indication of the crop yield loss can be obtained from the percentage of mildew infection present on leaf 3 at complete ear emergence of the crop (growth stage 59):

$$Y_L = 0.335\ 8 \times mi + 8.392\ 8 \qquad \ldots [5.5]$$

Yield loss, kg/ha $\propto$ Mildew, %

Whilst there is some correlation between the level of mildew at a particular stage of crop growth and yield, the fungal population dynamics depend on other variables such as weather, suscepti-

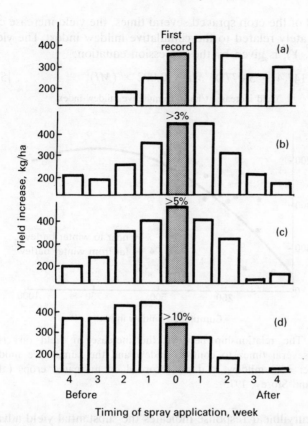

Fig. 5.6 The increase in yield associated with single sprays applied at the first record of mildew in the crop and when 3, 5 and 10 per cent mildew were first recorded on leaf 3 or leaf 4 and the increase in yields associated with single sprays applied up to 4 weeks before and after those application times (after Jenkins and Storey, 1975).

bility of the cereal variety, sources of inoculum, geographical situation, etc. The cumulative mildew index appears to be a better integrator of these factors. This index is obtained by measuring the area (per cent mildew × days) under the curve representing the progress of mildew on leaf 3 against time from the first record until complete ear emergence of the untreated crop. Taking the yield in the absence of mildew as equivalent to the yield of the crop sprayed several times, the yield increase can be accurately related to the cumulative mildew index. The yield increase, Y_I, is given by the regression equation:

$$Y_I = 147.4 + 3.077 \times MI - 0.002 \times (MI)^2 \qquad \dots [5.6]$$

Yield increase, t/ha $\propto$ Cumulative mildew index

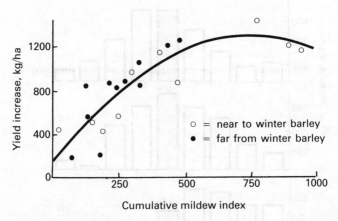

Fig. 5.7 The relationship between the increase in yield on crops sprayed several times to control mildew and the cumulative mildew index (per cent mildew × days) recorded on untreated crops (after Jenkins and Storey, 1975).

This curvilinear response indicates the substantial yield advantage by spraying to maintain a low cumulative mildew index in the treated crop, irrespective of whether the treated crop is adjacent to or remote from winter barley which can act as a reservoir for infection from the previous season (Fig. 5.7). The incremental yield advantage declines as the level of infection spreads to the point beyond which very severe or very early mildew infections depress the crop yield, presumably because even the multiple spraying programme is not totally effective in realising the full potential from the crop.

5.4 GRAIN LOSS DURING THE CEREAL HARVEST

There are two main sources of grain losses during the cereal harvest, those influenced by *harvest duration*:

○ **dry matter (respiratory) loss;**

○ **shedding loss;**
○ **cutter-bar (header) loss;**
○ **gale loss;**

and that influenced by *crop flow rate* through the harvester:

○ **separation loss.**

Dry matter and shedding losses are usually taken together, as both occur naturally before harvesting takes place. Cutter-bar loss is induced by the passage of the combine harvester through the crop. All duration losses increase with the time that the mature crop stands in the field before being cut. There are, however, significant differences in these losses for different crops and even for different varieties of the same crop. As barley straw is more brittle than wheat straw, the greater proportion of shedding and cutter-bar losses consists of broken-off complete ears in barley, and of loose grains in wheat. Thus, shedding accounts for the predominant loss in wheat, whilst cutter-bar losses are more important in barley.

For the assessment of harvest duration losses, a single ripening date is often adopted for the whole area whereas, in practice, a *spread of ripening* of less than seven days is unusual for a single cereal crop, such as spring barley. Where several cereal crops are grown in the rotation, e.g. winter wheat, spring barley and oats, a spread of ripening of up to four weeks is quite normal in Scotland due to the added effect of a range in altitude and/or mixture of light and heavy land.

An excessive spread of ripening can introduce the additional penalty of *gale losses* which can reduce yields by at least 25 per cent in affected crops.

Each size of combine harvester also has a design rate of work. Above this operating point, a higher crop flow rate overloads the grain separating mechanism and incurs a rapidly increasing *separation losses*.

The accurate assessment and interaction of the individual sources of grain loss influence the optimum size of a combine harvester and the alternative speed strategies for various crops and conditions on a particular farm.

5.4.1 Grain losses influenced by harvest duration

There are a number of equations which relate the percentage

shedding and cutter-bar losses in spring barley, y_s, to the number of days past crop ripeness for combine harvesting, X_2, a typical example being:

$$y_s = 2.2 \times e^{(0.05 \times X2)} \qquad \qquad \dots [5.7]$$

Shedding loss, % ∝ exp(Harvest duration function)

At the time of crop ripeness for combine harvesting, this equation predicts a negligible grain loss (Fig. 5.8) which is in agreement with the consistently low levels of loss found experimentally. The slope of the predicted loss curve is more arbitrary because of the much greater variability of the field data as the duration of the harvest is extended.

As the duration of the harvest encroaches on the autumn equinox, there is a greater risk of strong winds. Losses resulting from gales can be particularly severe but, fortunately, their frequency of occurrence is very low during the most susceptible period when the crop is fully ripe. A gale is defined as a period

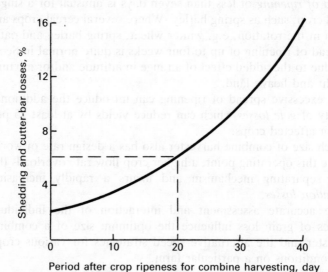

Period after crop ripeness for combine harvesting, day

Fig. 5.8 The effect of harvest duration on shedding and cutter-bar losses, as a percentage of the original yield at the time of crop ripeness for combine harvesting (after Elrick, 1982).

Table 5.5 Occurrence of gales at Edinburgh Airport during 70 days
commencing 1 August for 27 years 1956–82

Occurrence of gales		
24 August 1957	16 September 1961	21 September 1969
7 September 1957	17 September 1961	28 September 1969
24 September 1958	11 August 1962	15 September 1978
8 October 1958	11 September 1963	16 September 1978
27 August 1961	6 September 1966	

(*Source*: McGechan, 1985)

of 10 minutes with an average wind speed of over 63 km/h (Force 8 on the Beaufort Scale). Gales recorded at Edinburgh Airport during the 70-day period starting on 1 August, over 27 years, are listed in Table 5.5. Only one gale occurred within the harvest period during the past 10 years. Ignoring the fact that the expected frequency of gales is not constant but is lower at the start of the harvest season, the average frequency is one gale every 1.93 years. The probability of a gale on any given day is, therefore, 0.007 4 (i.e. 1/1.93 × 1/70). In addition, the greater predominance of earlier ripening winter cereals decreases the dangers still further.

The field observations of gale damage are incorporated in the gale loss curve in Fig. 5.9. There is evidence to suggest that crops at the unripe stage are immune from damage. Even when the crop just reaches the stage of ripeness for combine harvesting, only a moderate loss of, say, 7 per cent is sustained from gales. Standing crops which have been ripe for some time, however, can suffer losses of at least 25 per cent of the original yield. A sharp rise in the gale loss is appropriate between day 0 and day 15 if the gale loss can be regarded as an accelerated form of shedding loss which is itself curvilinear (see eqn [5.7]). After day 15, the curve is allowed to flatten out, since it is expected that increasing susceptibility would be offset by the protection gained when the ear breaks over with ripeness. It is this characteristic which is responsible for the rapidly increasing header losses in the later stages of crop maturity, already catered for in Fig. 5.8. The percentage loss in crop yield due to gale damage, y_g, is related to the number of days past crop ripeness for combine harvesting, X_2, by the equation:

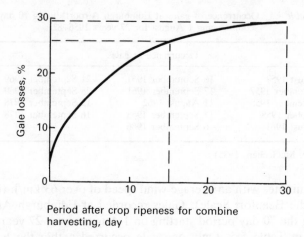

Fig. 5.9 Percentage loss in crop yield due to gale damage after the crop has reached the stage of ripeness for combine harvesting (after Elrick, 1982).

$$y_g = \frac{32 \times (X_2 + 5.2)^2}{100 + (X_2 + 5.2)^2} \qquad \ldots \text{[5.8]}$$

Gale loss, % $\propto$ Harvest duration, day

The percentage of total yield lost is further influenced, however, by the proportion of the crop at risk as the harvest proceeds, depending on the spread of ripeness and the daily rate of harvesting. For strategic planning purposes, it is reasonable to assume that ripening proceeds at a uniform rate so that if the spread of ripening, X_2, extends over 10 days, 1/10 of the crop area to be harvested ripens daily (Fig. 5.10); similarly, if harvesting proceeds at a uniform rate over a harvest duration, X_h, of 20 days, then 1/20 of the crop area is harvested daily. Provided that the harvest starts on the day when the first part of the crop reaches ripeness for combine harvesting, consecutive parts of crop are at risk for the timespan $(x_h - x_r)$ in Fig. 5.10, where x_h is the date of harvesting that part of the crop which ripens on day x_r. This timespan gives the period after crop ripeness for combine harvesting, X_2, which is required to calculate the shedding and cutter-bar losses from eqn [5.7]; it also gives the period during which a particular part of the crop is at risk

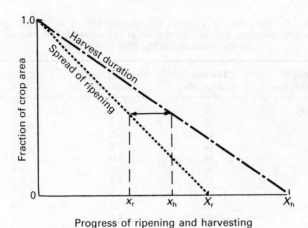

Progress of ripening and harvesting

Fig. 5.10 The timespan, for any fraction of a crop, between the date of harvesting, x_h, and ripening, x_r, as work proceeds (after Elrick, 1982).

from gales. As both shedding loss and gale loss are not linear with respect to time, the calculations of these losses for various harvest durations and spreads of ripening are presented in Table 5.6.

5.4.2 Separation loss

The separation loss rises with throughput of the combine harvester. At low throughputs, the loss rises slowly but, once throughput has reached the design capacity of the machine, further increases overload the grain/straw separating mechanism and cause a very large separation loss.

The rated throughput of a combine harvester, C_t, as quoted by the manufacturer, is usually based on tests in a crop of wheat with a grain:straw ratio of unity and a threshing loss of 2 per cent. In many crops, the grain:straw ratio, R_{gs}, is not equal to unity and it is more appropriate to relate threshing losses to the throughput of *material-other-than-grain* (m.o.g.) which is given by the equation:

$$C_a = 0.1 \times V \times W_e \times Y/R_{gs} \qquad \ldots [5.9]$$

Actual throughput, t/h = 0.1 × Operating speed, km/h × Effective operating width, m × Grain yield, t/ha ÷ Grain:straw ratio

Table 5.6 The effect of spread of ripening and harvest duration on grain losses from shedding, etc. and from gales as percentages of the original yield for the total crop in the East of Scotland

Spread of ripening, day	Harvest duration, day	Shedding, etc. losses, %	Gale losses, %
8	10	2.3	0.1
	12	2.5	0.3
	14	2.6	0.4
	16	2.8	0.5
	18	3.0	0.7
	20	3.2	0.9
	22	3.3	1.0
	24	3.5	1.2
	26	3.8	1.4
	28	4.1	1.6
	30	4.3	1.8
	32	4.6	2.0
	34	4.9	2.2
	36	5.3	2.5
	38	5.6	2.7
	40	6.0	2.9
15	16	2.3	0.1
	18	2.4	0.2
	20	2.5	0.3
	22	2.7	0.4
	24	2.8	0.5
	26	3.0	0.7
	28	3.2	0.9
	30	3.4	1.0
	32	3.6	1.2
	34	3.8	1.4
	36	4.0	1.6
	38	4.3	1.8
	40	4.6	2.0
22	24	2.3	0.1
	26	2.4	0.2
	28	2.6	0.3
	30	2.7	0.5
	32	2.9	0.6
	34	3.1	0.8
	36	3.2	0.9
	38	3.4	1.1
	40	3.6	1.3

(*Source*: Elrick, 1982)

The convex curve of separation loss against throughput is related to the ratio of the actual throughput, C_a, and the rated throughput, C_r, such that:

$$y_t = 2 \times [C_a/C_r]^2 \qquad \ldots [5.10]$$

$$\text{Separation loss, \%} = 2 \times \left[\frac{\text{Actual throughput, t/h}}{\text{Rated throughput, t/h}} \right]^2$$

For a combine harvester with a rated throughput of 9 t/h and a cutting width of 4 m, the effect of forward speed on the separation loss in a crop yielding 5 t/ha with a grain:straw ratio of unity is shown in Fig. 5.11. The separation loss at the rated throughput which is reached at a forward speed of 4.5 km/h doubles when the speed is increased to 6.4 km/h. In a lighter crop, (e.g. 4 t/ha m.o.g.), operating speeds can be increased

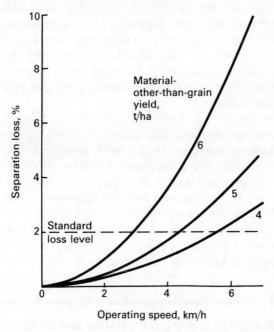

Fig. 5.11 The effect of operating speed and throughput on the separation loss for a combine harvester with a rated throughput of 9 t/h and a cutting width of 4 m.

whereas, in a heavy crop (e.g. 6 t/ha m.o.g.), the separation loss is very sensitive to speed.

5.5 VALUATION OF FORAGE LOSSES

Unlike cash crops which have an identifiable market price, the indirect costs associated with crops grown for home consumption are more difficult to assess because they involve the manipulation of livestock performance data. There are two types of indirect costs in forage conservation:

○ **losses in feed value of herbage with harvest duration;**
○ **conservation and storage losses.**

The first of these indirect costs refers to the failure to cut the grass at the target quality and the penalty becomes increasingly severe as harvest is prolonged. This has an immediate impact on the harvesting system because the addition of the duration penalty costs to the machinery operating costs effectively prescribes the optimum area capability of the equipment. On the other hand, the conservation and storage losses (which are extremely variable) are best considered separately to demonstrate the relative significance, at varying levels, compared with the combined effect of machine operating costs and penalties of untimely harvesting.

The valuation of the duration penalty losses depends on the saleable livestock product, different results being obtained for a beef production enterprise or for milk production. A further complication arises with a significant decline in nutritive value of the grass. Supplementation of the ration is necessary to achieve a realistic level of production but also depresses intake of the conserved grass. This supplementation can be at low cost by replacing part of the forage area with other home-grown crops, such as roots or pulses for energy and protein, respectively, or at high cost with bought-in concentrates.

5.5.1 Harvest duration penalties with grass for silage

Typical changes in the dry matter yield and in the digestible organic dry matter (DOMD) for first and second cuts are shown in Fig. 5.12 for a pure S24 perennial ryegrass sward. The energy concentration of the resulting herbage is given by the formula:

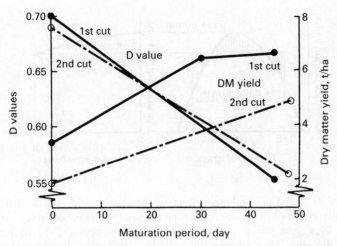

Fig. 5.12 Dry matter yield and D values for first and second cuts of S24 perennial ryegrass (after Green *et al.*, 1971).

$$ME = 0.82 \times D_v \times E \qquad \ldots [5.11]$$

Metabolisable energy, MJ/kg = 0.82 × Digestible organic dry matter, dim.
× Gross energy value of DOMD, MJ/kg

where 0.82 is the factor for deriving metabolisable energy from digestible energy, and the gross energy value of the digestible organic dry matter is taken as 20 MJ/kg for direct cut grass and 19 MJ/kg for wilted grass to account for additional field losses associated with wilting. The metabolisable energy production per hectare can then be derived for grass cut at D values ranging from 0.70 to 0.55 and from 0.68 to 0.56 for first and second cut material, respectively, assuming that the grass cut and silage fed are of identical value, i.e. no losses between cutting and feeding, as these losses are considered separately.

Using beef as the saleable livestock product, the liveweight gains per hectare from each level of energy production depend on the ration formulation, animal liveweight, grass quality and silage dry matter. For a sole diet of silage, fed to appetite, the liveweight increase responses are different for various ages of animals. The dry matter intake levels for two liveweights of animal, 300 kg and 450 kg, and two silage dry matters, 20 per

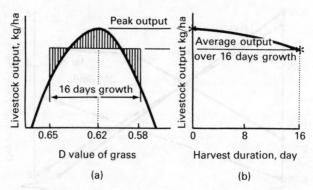

Fig. 5.13 (a) The combined effect of increasing yield and decreasing quality as grass matures by converting the available metabolisable energy into liveweight gain at different D values of the herbage; (b) assuming that the grass is not all mown at one time, the average livestock output is the mean of the livestock outputs for the different D values of the grass spanned by the harvest period.

cent and 30 per cent (representing wilted and direct cut silage), are based on the intake equation for silage with a D value of 0.64:

$$I_{L1} = 0.016 \times W_L + 0.042 \times DM - 0.37 \qquad \ldots [5.12a]$$

Intake, kg $\propto$ Animal liveweight, kg + Silage dry matter, %

The intake is further adjusted by 0.1 kg for each unit change in D value increasing above and decreasing below the 0.64 D value intake level for both steer liveweights. The livestock outputs which the silage diets of varying qualities can produce are shown in Fig. 5.13(a). In practice, harvesting takes place over a period of time so the D value changes as the harvest progresses. For this reason, it is better to think in terms of *average* livestock output with the harvest period extended equally on each side of the D value which would achieve *peak* livestock output (Fig. 5.13(b)). The fall in average output gives the duration penalty cost.

Although no allowance is made for changes in crude protein content which occur with declining D values, a revised form of the intake equation for beef production from weaned suckled calves now includes the effect of changing D value and level of ammonia-nitrogen of the silage:

$$I_{L2} = 0.007\,5 \times W_L + 0.105 \times DM + 15.6 \times D_v - 0.02 \times N_A$$
$$+ 3.5 \qquad \qquad \dots [5.12b]$$

Intake, g/kg LWt $\propto$ Animal liveweight, kg + Silage dry matter, % + Digestible organic dry matter, dim − Ammonia nitrogen, g/kg total N

Other intake equations account for the changes in silage intake with mixed rations and for different classes of stock.

The average duration penalty costs are represented by the decrement in average livestock output with harvest duration. As these penalty costs occur irrespective of whether the grass is ensiled fresh or wilted, the average losses as a percentage of peak livestock output only vary with liveweight and with number of cuts, the losses being higher for the lighter steers and for fewer cuts (Fig. 5.14). The heavier beef cattle produce their peak output at a lower D value than for the lighter animals and are less sensitive to declining quality. There is, of course, the prac-

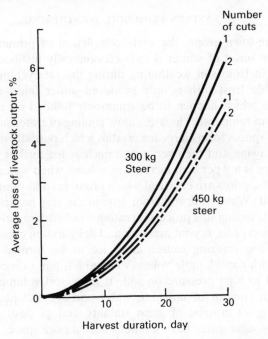

Fig. 5.14 Average losses of livestock output for one and two cuts of ensiled grass and two steer liveweights.

tical difficulty of determining when to start harvest, and the risk of incurring higher penalty costs when longer harvests are spread about a target D value.

5.2.2   Feed losses during grass silage conservation and storage

Losses subsequent to cutting grass occur for a number of different reasons and to varying degrees. The silage mechanisation system can be responsible for at least a proportion of such losses; for example, relating to the method of picking up the grass, the speed of filling the silo and the ease and success of consolidation. As a comparison with the duration penalty losses, the losses from cutting to feeding, in terms of livestock output forgone, range from 10 per cent for a wilted silage to 20 per cent for direct-cut silage.

5.6   SAVINGS FROM SOIL WEATHERING

For spring-sown crops, the early completion of primary tillage before the onset of winter is only economically justified if there is a benefit from soil weathering during the intervening period. A desirable frost tilth is only produced under fairly restricted conditions which happen to be commonly fulfilled on loam and clay soils in temperate climates. Slow cooling of water in tension-free pore spaces forms large ice crystals which partially melt away during thawing and then serve as a nucleus for further freezing. Since there is a 9 per cent increase in volume when water changes into ice, the pores are enlarged which promotes a loosening effect on the soil. Water is drawn from around the clay particles to the ice crystals during this process, creating a dehydration effect. The combination of ice crystal pressure and dehydration causes aggregation. Slow freezing causes more ice to be formed in thick wedges with considerable volumes of soil left almost ice-free and subjected to high pressure on soil. If these soil volumes have a fairly high content of unfrozen water and are still plastic, this pressure gives crumbs of great stability and of desirable size. These new aggregates have smaller internal pore spaces than the original clods, owing to the pressure exerted on them, so that when the soil thaws, much of the ice will become free water and drain away. It is by this means that the effect of slow freezing and thawing of wet clods is to leave the soil drier than it was before the frost with a more stable crumb structure of a desirable size and distribution.

The various factors affecting soil aggregate breakdown, *BD*, have been combined in a regression equation:

$$BD = -55.6 \times \theta_t - 25.2/N_0 - 2.03 \times T_s + 0.76 \times Si \\ - 3.19 \qquad \qquad \cdots [5.13]$$

Soil aggregate breakdown, % ∝ − Soil moisture tension, atmos − Number of freezing and thawing cycles − Soil temperature, °C + Silt content, %

The negative moisture tension coefficient implies that an increase in soil moisture content causes an increase in aggregate breakdown. There is also a more destructive effect on soil aggregation with a greater number of freezing and thawing cycles at lower temperatures. This equation forms the basis for analysing weather records to assess the number of tillage work days which would require to be sacrificed for a given level of aggregate breakdown due to natural weathering. The associated increase in primary tillage costs for early completion can then be offset by the financial benefit of this aggregate breakdown in the form of one or two less secondary cultivations.

5.7 SUMMARY

Indirect machinery charges are the penalties (or, occasionally, savings) incurred by scheduling machinery operations over a period of time.

Balanced scheduling is when machinery operations are spread equally on either side of the day for maximum crop response; premature scheduling and late scheduling are optional operating strategies.

Crop yield losses through untimely establishment are proportional to the square of the time deviation from the optimum date of establishment.

Crop yield losses from delayed planting can be related to the work rate of crop establishment machinery for planning system requirements.

Grain losses during harvest are influenced by both harvest duration and crop flow rate through the harvester but can be minimised by careful choice of crop varieties to extend the spread of ripening.

Harvest duration penalties related to forage conservation involve the manipulation of livestock performance data to provide a market value for the livestock output forgone.

Early completion of primary tillage for spring-sown crops can reduce the number of secondary cultivations by providing more opportunity for soil weathering processes.

6

SOIL, WEATHER AND WORKDAYS

OUTLINE

Soil moisture budget; Runoff; Hydraulic conductivity; Drainage; Evapotranspiration; Soil workability; Soil workdays; Spraying occasions; Spray drift hazard; Forage drying in the swath; Equilibrium moisture content; Cumulative vapour pressure deficit; Sequences of dry days; Morphological and technological maturity of cereals; Field moisture content of standing grain; Criteria for combine harvesting.

APOSTROPHE – BY HOOK OR BY CROOK

From the laws of Canute issued in Winchester in 1016, and from subsequent entries in the Domesday Book, it is evident that much of the area of the present New Forest was already the property of the Crown. Although much of the land in the Royal Forest was probably transferred from Monastic Orders, numerous small pockets of an acre or more remained in private hands and the owners were subject to the severe Norman Forest Laws drawn up for the sole protection of the deer. It is thought that those very minimal privileges are the origins of the 'Common Rights on the Forest' which include the Common of Pasture to graze stock, the Common of Mast to turn out pigs during the pannage season, the Common of Turbary to cut peat for burning, the Common of Marl to manure the land, and the Common of Fuelwood. Of these ownership privileges, the Common of Pasture is the only Right which is exercised at the present time, whilst the Common of Fuelwood has long been resisted by the Crown and, wherever possible, has been either extinguished or commuted to the free provision of a few loads of firewood each year.

Nevertheless, for centuries, the New Forest Commoners have

fought to preserve the Right of *Estovers*, a Norman-French
word related to our 'stove' which implied the collection of fuelwood. The Commoners could gather wood 'by hook or by
crook'. The crook, a bent stick used to dislodge dead branch-
wood, did little damage; but the 'hook' was often a billhook
used to sever any stem not large enough to need an axe.

Now, if a young broad-leaved tree is cut back in this way,
it gives rise to a cluster of *coppice* shoots – another Norman-
French word derived from couper, to cut. These shoots could
be harvested in turn, but they were more likely to be browsed
by the Commoners' ponies and cattle.

The keepers of the Royal Forest much preferred the practice
of pollarding, a method named from the old word 'poll' for
head, for the trees were beheaded at a height of 6 feet or so
to produce a cluster of branches safely above the reach of cattle
and deer. The annual growth from the pollarded trees was used
by the Keepers to provide 'browse' or winter keep for the deer;
and, more important, once the deer had nibbled the bark, the
Keepers were allowed to sell the remainder for firewood and
retain the proceeds!

Hence, the proverb 'by hook or by crook' has the dual
meaning of 'rightfully or wrongfully' and alternatively 'in one
way or another', the latter interpretation being more appropriate in the context of getting the job done.

(Sources: Pigott 1960; Edlin 1960)

6.1 INTRODUCTION

Soil moisture content and the weather are the two major factors
which determine the amount of time available throughout the
year for field operations. In a poor season, little time may be
available for performing one or several field operations under
acceptable conditions. A favourable weather pattern and a firm
soil, on the other hand, provide abundant time in which the field
work can be completed without working excessively long hours,
or working in unsatisfactory conditions. The wide variation in the
time available from year to year creates a major management
problem in correctly sizing the machinery complement. One
management approach is to invest heavily in machinery to
combat the most adverse conditions which results in substantial
spare machine capacity under average conditions. The alternative
management strategy is to maximise returns over a period of

years by balancing the cost of modest over-investment in machinery for the favourable years with the benefit of more timely operations during the least favourable seasons. Whilst the latter approach is more profitable, it requires not only historical data on the duration and frequency of recurrence of workday periods which satisfy the operational criteria but also probabilities for forward planning with weather uncertainty.

The weather interacts both with the soil to vary soil workability for tillage operations and with the crop to vary yield and moisture content at maturity for harvesting operations, whilst the influence of the weather on soil trafficability affects all operations to a greater or lesser extent. The workability and trafficability of the soil is dependent on the soil moisture content which can be evaluated from soil and weather variables. For crop spraying, low wind speeds are preferred to minimise spray drift during the operation itself, but a rather longer rain-free period is necessary to ensure that the crop canopy is dry enough to receive the spray and to provide sufficient time for the active ingredients to develop their rain-fast properties. Forage and grain quality may be enhanced and drying costs are reduced by harvesting in airy conditions when rapid crop moisture loss can be achieved in the field through a combination of wind and low relative humidity.

Provided that all the relevant operating conditions can be specified for the soil and the crop, suitable workdays can be identified. There remains, however, the additional management factor which allows the workday criteria to become a moving target. The definition changes depending on whether the field activities are well advanced or are already delayed in relation to the average timespan anticipated. If the work is behind hand, it is necessary to operate in poorer conditions to salvage the situation. Day-to-day operating tactics which vary the workday criteria are similar to moving the goal posts – it is a different ball game which is based on a judgement of the economic viability of the enterprise as a whole (see section 9). As a preliminary stage, however, commonly accepted operational windows are used to provide workday data for strategic planning purposes.

6.2 SOIL MOISTURE CONTENT

Rainfall alone is of little help in determining the moisture content of the soil because drainage of water through the soil varies markedly from site to site. Both evaporation of moisture from the soil and transpiration of water vapour by the crop are also affected by the ambient temperature, with warmer intervals favouring a more rapid moisture loss. In addition, the existing moisture status of the soil governs the rate of drainage and the rate of evapotranspiration. As the soil dries out, the removal of additional water becomes progressively more difficult. Not accounting for this fact in the past has led to excessively high estimates of soil workdays. Lastly, during periods of high intensity or prolonged rainfall, the percolation of water through the soil is too slow to meet the deluge imposed upon it and the excess flows over the soil surface to the nearest waterway as runoff.

In order to evaluate the soil moisture content on a daily basis, it is necessary to consider:

○ **soil moisture content on the previous day;**
○ **precipitation;**
○ **runoff;**
○ **drainage;**
○ **evapotranspiration.**

These factors are combined to form the soil moisture balance equation for the top 300 mm of the plough layer:

$$m_a = m_p + Q_p - Q_r - Q_d - Q_e \qquad \ldots [6.1]$$

Soil moisture content, mm = Soil moisture content on previous day, mm + Precipitation, mm − Runoff, mm − Drainage, mm − Evapotranspiration, mm

By means of this equation, the soil moisture content on the previous day is adjusted for soil moisture gains and losses to provide a revised soil moisture content on a cumulative basis.

For simplicity, only water entering the soil as precipitation (snow, sleet, rainfall and/or irrigation water) need be considered because worldwide meteorological data are readily available. In temperate climates, the amount of water condensing onto the soil from the air entering the soil may be neglected, although this moisture source is important in some parts of the world where dry farming is practised. Equally, the transfer of water from adjacent areas either as surface runoff or subsurface flow only can be ignored for non-sloping sites and homogeneous soil profiles. The differential transmission of water vapour through the soil is negligible in comparison with the evapotranspiration and, at the early stages of plant establishment, the moisture directly intercepted by vegetation can be discounted. Despite these provisos, the moisture content can be predicted with a high degree of accuracy.

6.2.1 Surface runoff

Surface runoff depends on the moisture-retention characteristics of the soil. The amount of moisture in the upper part of the soil profile immediately prior to precipitation provides a good *index of antecedent soil moisture* with which to estimate the water retained from any particular amount of rainfall.

- The **antecedent soil moisture index** is the water storage potential of the soil profile after deducting the soil water held at the *permanent wilting point*.

This index is calculated from the equation:

$$I_m = m_p - PWP \times h \times \gamma_d/10 \qquad \ldots [6.2]$$

Antecedent soil moisture index, mm = Soil moisture on previous day, mm
− Permanent wilting point, % w/w × Depth of soil profile, mm
× Soil bulk density, kg/m³ ÷ 10

The permanent wilting point is used to provide a practical baseline for the available water reservoir capacity, and values typically range from 15 to 20 per cent w/w.

- The **permanent wilting point** is the soil moisture content at 15 atmospheres of soil water tension, beyond which the plant cannot effectively obtain water.

Any soil moisture content can be converted into an equivalent depth of water by multiplying its decimal value by the depth of the soil profile and the ratio of the soil bulk density to the density of water (1000 kg/m³).

The relationship between the daily rainfall, Q_p, and the proportion retained by the soil $(Q_p - Q_r)/Q_p$ takes the form:

$$Q_p/(Q_p-Q_r) = a_1 + Q_p/b_1 \qquad \ldots [6.2a]$$

1/(Rainfall proportion retained by soil, dec.) = Intercept constant, dim
+ Rainfall, mm ÷ Slope constant, mm

The most important factor derived from the runoff equation (eqn [6.2a]) is that the amount of rainfall, Q_{po}, which could be retained by the soil with no runoff (i.e. when $Q_r = 0$) is such that:

$$a_1 = 1 - Q_{po}/b_1 \qquad \ldots [6.3]$$

Intercept constant, dim = 1 − Rainfall at zero runoff, mm ÷ Slope constant, mm

Substituting for the intercept constant, the runoff equation becomes:

$$Q_r = Q_p - \frac{b_1 \times Q_p}{b_1 + Q_p - Q_{po}} \qquad \ldots [6.2b]$$

From an analysis of experimental data, the amount of rainfall which could be retained with no runoff and the slope constant are both influenced by the antecedent soil moisture index. Provided that this index does not exceed a value of 200 mm for the water storage capacity in the topsoil layer of 300 mm, the daily runoff is:

$$Q_r = Q_p - \frac{Q_p(615 - 2.85 \times I_m)}{Q_p + 529 - 2.44 \times I_m} \qquad \ldots [6.4]$$

Runoff, mm ∝ Rainfall, mm and Antecedent soil moisture index, mm

More runoff occurs on wetter soils with higher values of the antecedent soil moisture index (Fig. 6.1).

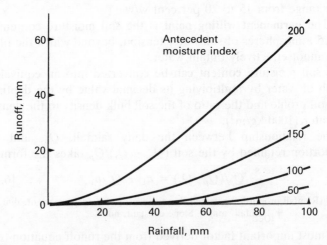

Fig. 6.1 The effect of rainfall and antecedent moisture index on runoff.

6.2.2 Drainage

Soil water movement is governed by the hydraulic conductivity. Whilst there are substantial fluctuations in the rate of soil water movement, various approximations are appropriate for *particular* applications but these approximations cannot be transferred from one application to another. For the design of field drainage systems, for example, it is only essential to identify the maximum flow rate for the selection of an adequate system capacity and it is quite acceptable to adopt the concept of a *drainage coefficient* (Fig. 6.2).

● The **drainage coefficient** indicates a constant daily rate of water movement draining through the soil profile.

This approach implies (incorrectly) that the rain water percolates steadily from the soil surface until it is completely removed at a lower level, after which the drainage flow drops instantaneously from a maximum rate to zero. The simple representation of drainage flow is appropriate for drainage design, but totally unacceptable in the evaluation of soil moisture changes.

A more accurate description of soil water movement is derived from experiments in 1857 by Darcy who found that the flow rate in porous materials is directly proportional to the *hydraulic gradient*, which is a measure of the pressure forcing the water to flow through the soil.

● The **hydraulic gradient** is the ratio of the soil water pressure head to the depth of the soil profile.

Darcy's law may be written as:

$$Q_d = K_{sat} \left\{ \frac{d\emptyset}{dh} \right\} + 1 \qquad \dots [6.5]$$

Drainage flow, mm/day = Hydraulic conductivity, mm/day
    × (Soil water pressure head, mm × Soil depth, mm) + 1

The proportionality factor, K_{sat}, is the *hydraulic conductivity*.

● The **hydraulic conductivity** is the rate at which water flows through the soil profile.

The hydraulic conductivity is constant for saturated soils but as the soil dries out, water moves primarily in small pores and through films located around and between the soil particles. With decreasing soil moisture content, the cross-sectional area of the water films is reduced and the water flow paths become more limited. In consequence, the hydraulic conductivity falls very rapidly with decreasing soil moisture content. In comparison with these very large changes, the variation in the hydraulic gradient in a uniform soil is negligible. When the drainage flow is plotted against the soil moisture content on logarithmic scales, the features of unsaturated flow can be expressed in a linear equation for drainage, of the form:

$$\ln Q_d = c_1 \times \ln m_p + c_2 \qquad \ldots [6.6]$$

ln Drainage flow, mm/day = Soil type constant 1 × ln Soil moisture content on previous day, mm/300 mm + Soil type constant 2

The hydraulic conductivity and hence the drainage rate reaches a constant upper limit when the soil is at saturation and either ceases or reaches a negligible value when the soil moisture content approaches *field capacity* (Fig. 6.2).

● **Field capacity** is the soil moisture content at 0.3 atmospheres of soil water tension beyond which drainage flow effectively ceases.

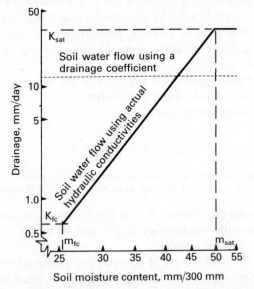

Fig. 6.2 Hydraulic conductivities, K, for the soil moisture contents, m, at saturation and field capacity used in the determination of drainage flow.

Table 6.1 Soil water properties for a range of soil types

Soil series (Soil type)	Soil bulk density kg/m³	Soil moisture content in 300 mm soil profile				Hydraulic conductivity in 300 mm soil profile, mm/day	
		Wilting point, % w/w	Field capacity, % w/w	Plastic limit, % w/w	Saturation, % w/w	Field capacity	Saturation
Darvel (Sandy loam)	1200	12.5	25.5	30.5	46.2	1.0	38.2
Macmerry (Medium loam)	1100	14.0	25.7	27.5	49.1	0.8	32.3
Winton (Clay loam)	1100	17.0	27.2	28.3	46.0	0.5	21.3

Solving eqn [6.6] for the upper and lower limits of hydraulic conductivity, K_{sat} and K_{fc}, when the soil moisture content is at saturation, m_{sat}, and at field capacity, m_{fc}, the drainage equation becomes:

$$Q_d = K_{fc} \times (m_p/m_{fc})^\alpha \qquad \ldots [6.7]$$

Drainage flow, mm/day = Hydraulic conductivity at field capacity, mm/day
    × (soil moisture content on previous day, mm/300 mm ÷ Soil moisture
    content at field capacity)$^{\text{Drainage rate exponent}}$

where:

$$\alpha = (\ln K_{sat} - \ln K_{fc})/(\ln m_{sat} - \ln m_{fc}) \qquad \ldots [6.8]$$

Drainage rate exponent = Difference of ln values of hydraulic conductivities at
    saturation and field capacity ÷ Difference of ln values of soil moisture
    content at saturation and field capacity

The major advantage of this relationship is that soil water movement can be accurately estimated from a knowledge of simple soil water properties. Typical values of these soil water properties are given in Table 6.1 for a range of soil types.

6.2.3   Evapotranspiration

As well as runoff and drainage, evapotranspiration causes major fluctuations in the soil moisture content. Evapotranspiration is a combination of two moisture loss mechanisms:

○ **evaporation;**
○ **transpiration.**

The rate or evaporation from the soil and from the leaves of any vegetation varies with the wetness of the surfaces. For this reason, the standard evaporation rate is often based on the rate for a free water surface.

● **Evaporation rate** is the rate of water loss from the liquid to the vapour phase from a free water surface.

The amount of transpiration by the plant also varies with the water availability in the root zone and with the stage of plant development.

● **Transpiration rate** is the rate of water loss through the stomata, or pores, in the leaves of plants.

Both evaporation and transpiration rates are additionally influenced by the ambient climatic conditions and complex formulae have been developed to determine the rate of evapotranspiration with a high degree of accuracy to facilitate the correct scheduling of irrigation water application. The greater the complexity of the evapotranspiration formulae, however, the greater the meteorological data required and this detailed information is often only available from a limited number of weather stations. For more widespread applicability, an empirical procedure has been adopted for the assessment of the actual evapotranspiration rate using correction factors to adjust the rate of potential evapotranspiration under optimum conditions.

The monthly potential evapotranspiration rate can be assessed from the mean monthly air temperatures, using the relation:

$$E_p = 16 \times (10 \times T_a/I_h)^\beta \qquad \ldots [6.9]$$

Potential evaporation rate, mm/mnth ∝ Mean monthly air temperature, °C

where

$$\beta = (0.675 \times I_h^3 - 77.1 \times I_h^2 + 17\,900 \times I_h + 492\,000) \times 10^{-6}$$
$$\ldots [6.10]$$

Evaporation rate exponent ∝ Annual heat index

and:

$$I_h = i_1 + i_2 + i_3 + \ldots + i_{12} \qquad \ldots [6.11]$$

Annual heat index = Sum of monthly heat indices

$$i_n = (T_{an}/5)^{1.514} \qquad \ldots [6.12]$$

Monthly heat index ∝ Mean monthly air temperature, °C

with the subscripts n from 1 to 12 defining the months of the year.

Since there is a linear relation between the logarithm of temperature and the logarithm of the unadjusted potential evapotranspiration rate, the variation in the rate of evapotranspiration at different locations is governed by the slope of the lines and this is determined from the effect of the heat index on the evaporation exponent (Fig. 6.3). All the lines converge at a temperature of 26.5 °C and at a potential evapotranspiration rate of 135 mm/month. For a cooler climate, the annual heat index is lower and the decline in evapotranspiration rate becomes more rapid with falling air temperatures.

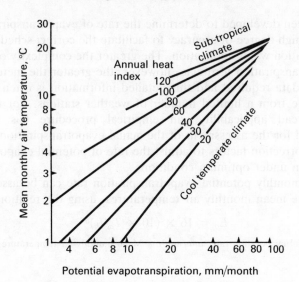

Fig. 6.3 The effect of the annual heat index on the rate of potential evapotranspiration.

The monthly evapotranspiration rate, calculated from eqn [6.9], is divided by the number of days (30) in a standard month to give the daily potential evapotranspiration rate and factorially adjusted to yield the actual evapotranspiration rate:

$$E_a = E_p \times K_l \times K_d \times K_w \times K_g \times K_s/30 \qquad \ldots [6.13]$$

Actual evapotranspiration rate, mm/day = Potential evapotranspiration rate, mm/month × Day length factor × Soil dryness factor × Wet days factor × Crop stage of growth factor × Surface cover factor ÷ 30

The *day length correction factor*, K_l, introduces the actual duration of sunshine, N_s, relative to the duration of daylight, N_l, for a particular latitude from the Smithsonian table (Table 6.2):

$$K_l = N_s/N_l \qquad \ldots [6.14]$$

Day length correction factor, dim = Number of sunshine hours, h/month ÷ Number of daylight hours, h/month

The *soil dryness correction factor*, K_d, is defined as the ratio of the actual and potential evapotranspiration rates. When the soil moisture content is near or at field capacity, the actual evapo-

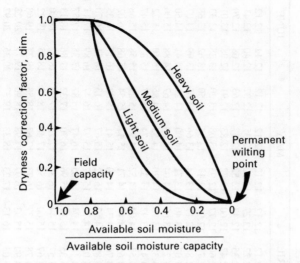

Fig. 6.4 The effect of soil moisture deficit on the soil dryness correction factor.

transpiration occurs at the potential rate if other conditions are optimum. As the soil moisture is depleted, so the actual evapotranspiration rate gradually declines to the potential rate at the permanent wilting point. The variation in the values of the dryness correction factor with the soil moisture deficit for a 300 mm profile is shown in Fig. 6.4.

The *wet day correction factor*, K_w, is designed to compensate for some of the shortcomings of the potential evapotranspiration rate formula based on temperature alone by making an allowance for variations due to cloudiness, humidity and the occurrence of precipitation. The appropriate values for the wet day correction factor are listed in Table 6.3.

The purpose of the *crop stage of growth factor* is to adjust for the progressive variation in evapotranspiration rate over a typical growing season. For grass, evapotranspiration at the potential rate is reached shortly before harvesting (Fig. 6.5). Removal of most of the transpiring foliage at that time immediately curtails the actual evapotranspiration rate to about 40 per cent of the potential rate, but the renewal of growth gradually restores the ability of the crop to transpire. Note, however, that the crop

Table 6.2 The effect of latitude and season on the duration of daylight (Smithsonian Meteorological Tables)

Duration of daylight, h

Lat. deg.	JANUARY			FEBRUARY			MARCH			APRIL			MAY			JUNE		
0	12.12	12.12	12.12	12.12	12.12	12.12	12.12	12.12	12.12	12.12	12.12	12.12	12.12	12.12	12.12	12.12	12.12	12.12
5	11.85	11.87	11.88	11.93	11.96	12.00	12.04	12.08	12.13	12.18	12.22	12.27	12.32	12.35	12.37	12.40	12.41	12.42
10	11.56	11.61	11.66	11.73	11.81	11.90	11.97	12.06	12.15	12.27	12.35	12.43	12.52	12.57	12.63	12.68	12.70	12.72
15	11.28	11.34	11.42	11.53	11.66	11.78	11.90	12.04	12.18	12.33	12.47	12.60	12.72	12.83	12.91	12.98	13.01	13.02
20	10.97	11.06	11.17	11.33	11.50	11.67	11.82	12.01	12.20	12.42	12.60	12.77	12.93	13.08	13.20	13.28	13.33	13.35
25	10.65	10.76	10.92	11.11	11.33	11.55	11.75	11.98	12.23	12.50	12.74	12.97	13.17	13.35	13.50	13.62	13.68	13.68
30	10.30	10.42	10.63	10.88	11.13	11.42	11.67	11.97	12.27	12.58	12.88	13.17	13.44	13.66	13.85	13.99	14.07	14.08
35	9.90	10.07	10.30	10.62	10.95	11.28	11.58	11.97	12.30	12.70	13.07	13.40	13.71	14.00	14.23	14.41	14.50	14.50
40	9.45	9.65	9.93	10.32	10.70	11.12	11.47	11.92	12.35	12.83	13.25	13.66	14.04	14.38	14.67	14.88	15.00	15.00
42	9.25	9.45	9.78	10.19	10.60	11.05	11.42	11.90	12.37	12.88	13.35	13.78	14.20	14.55	14.85	15.10	15.22	15.23
44	9.02	9.25	9.60	10.04	10.49	10.97	11.37	11.86	12.39	12.94	13.43	13.92	14.35	14.74	15.07	15.33	15.47	15.47
46	8.78	9.05	9.42	9.89	10.38	10.90	11.32	11.87	12.42	13.00	13.53	14.04	14.52	14.95	15.29	15.58	15.68	15.75
48	8.54	8.82	9.21	9.73	10.25	10.82	11.27	11.85	12.43	13.07	13.65	14.19	14.71	15.17	15.54	15.68	16.02	16.03
50	8.26	8.57	8.99	9.55	10.12	10.72	11.21	11.83	12.47	13.16	13.76	14.34	14.90	15.41	15.82	16.17	16.35	16.35
52	7.95	8.27	8.75	9.37	9.97	10.61	11.14	11.83	12.49	13.22	13.89	14.52	15.12	15.67	16.13	16.51	16.70	16.72
54	7.60	7.96	8.47	9.14	9.81	10.52	11.08	11.80	12.53	13.32	14.02	14.72	15.37	15.97	16.47	16.89	17.11	17.12
56	7.12	7.61	8.17	8.91	9.63	10.39	11.00	11.78	12.57	13.42	14.18	14.94	15.65	16.27	16.87	17.35	17.57	17.60
58	6.74	7.21	7.80	8.63	9.42	10.25	10.92	11.77	12.60	13.52	14.35	15.17	15.96	16.70	17.32	17.86	18.13	18.15
60	6.22	6.72	7.44	8.33	9.20	10.10	10.82	11.73	12.66	13.65	14.57	15.44	16.31	17.12	17.96	18.47	18.80	18.83
61	5.90	6.46	7.22	8.16	9.07	10.02	10.77	11.72	12.67	13.71	14.67	15.60	16.52	17.40	18.17	18.85	19.22	19.22
62	5.53	6.16	6.97	7.98	8.93	9.93	10.72	11.71	12.70	13.78	14.78	15.77	16.73	17.67	18.53	19.27	19.68	19.69
63	5.13	5.82	6.70	7.78	8.80	9.83	10.65	11.68	12.73	13.87	14.90	15.93	16.97	17.98	18.90	19.77	20.22	20.25
64	4.68	5.45	6.40	7.55	8.63	9.73	10.60	11.67	12.77	13.95	15.03	16.13	17.23	18.33	19.37	20.35	20.92	20.98
65	4.15	5.05	6.08	7.32	8.47	9.62	10.53	11.67	12.80	14.03	15.18	16.35	17.53	18.72	19.88	21.08	21.87	21.97

Table 6.2 Continued

Duration of daylight, h

Lat., deg.	JULY			AUGUST			SEPTEMBER			OCTOBER			NOVEMBER			DECEMBER		
0	12.12	12.12	12.12	12.10	12.11	12.10	12.11	12.11	12.11	12.12	12.12	12.12	12.12	12.12	12.12	12.13	12.12	12.12
5	12.36	12.38	12.40	12.28	12.33	12.23	12.10	12.16	12.19	11.97	12.02	12.05	11.87	11.90	11.92	11.83	11.83	11.85
10	12.60	12.64	12.68	12.45	12.53	12.37	12.08	12.17	12.27	11.83	11.91	12.00	11.61	11.67	11.75	11.54	11.54	11.58
15	12.85	12.94	12.98	12.62	12.73	12.50	12.07	12.22	12.35	11.67	11.82	11.95	11.35	11.43	11.55	11.24	11.22	11.28
20	13.12	13.23	13.30	12.83	12.97	12.62	12.07	12.25	12.45	11.52	11.70	11.88	11.07	11.19	11.35	10.93	10.93	10.97
25	13.41	13.56	13.63	13.02	13.22	12.82	12.05	12.31	12.55	11.35	11.58	11.82	10.78	10.93	11.12	10.60	10.60	10.67
30	13.72	13.89	14.02	13.23	13.49	12.96	12.07	12.34	12.65	11.18	11.47	11.76	10.47	10.65	10.89	10.22	10.23	10.32
35	14.07	14.29	14.45	13.48	13.78	13.16	12.06	12.42	12.77	10.98	11.33	11.68	10.10	10.34	10.63	9.82	9.82	9.93
40	14.48	14.74	14.92	13.77	14.13	13.37	12.06	12.48	12.91	10.77	11.18	11.60	9.70	9.88	10.33	9.35	9.35	9.84
42	14.67	14.92	15.12	13.90	14.29	13.46	12.03	12.51	12.97	10.67	11.12	11.57	9.52	9.83	10.22	9.15	9.15	9.29
44	14.87	15.17	15.37	14.02	14.44	13.57	12.04	12.54	13.03	10.57	11.04	11.54	9.32	9.67	10.07	8.90	8.90	9.00
46	15.07	15.40	15.62	14.17	14.62	13.68	12.03	12.57	13.11	10.45	10.97	11.51	9.11	9.48	9.92	8.65	8.68	8.83
48	15.30	15.66	15.91	14.32	14.82	13.80	12.03	12.61	13.18	10.33	10.88	11.46	8.86	9.28	9.75	8.37	8.41	8.57
50	15.55	15.95	16.22	14.50	15.03	13.92	12.02	12.67	13.27	10.21	10.80	11.41	8.63	9.07	9.58	8.09	8.12	8.30
52	15.83	16.27	16.56	14.67	15.26	14.06	12.02	12.69	13.35	10.07	10.70	11.37	8.37	8.82	9.37	7.77	7.82	8.00
54	16.15	16.63	16.95	14.88	15.52	14.21	12.02	12.74	13.44	9.91	10.60	11.32	8.05	8.56	9.17	7.40	7.42	7.66
56	16.50	17.04	17.40	15.12	15.81	14.38	12.02	12.81	13.55	9.74	10.48	11.26	7.71	8.26	8.95	6.97	7.00	7.26
58	16.91	17.52	17.92	15.38	16.15	14.57	12.02	12.85	13.67	9.54	10.37	11.20	7.32	7.93	8.68	6.48	6.52	6.81
60	17.38	18.07	18.57	15.67	16.52	14.81	12.02	12.92	13.81	9.33	10.22	11.12	6.85	7.57	8.38	5.90	5.93	6.28
61	17.67	18.40	18.95	15.83	16.73	14.91	11.98	12.93	13.89	9.22	10.15	11.07	6.60	7.34	8.21	5.57	5.61	5.98
62	17.98	18.77	19.40	16.02	16.97	15.05	12.00	13.00	13.97	9.08	10.05	11.03	6.30	7.12	8.03	5.18	5.22	5.62
63	18.30	19.19	19.90	16.21	17.23	15.25	12.00	13.03	14.05	8.95	9.97	11.00	5.97	6.87	7.83	4.73	4.80	5.25
64	18.68	19.67	20.52	16.42	17.50	15.33	12.00	13.08	14.15	8.80	9.87	10.95	5.62	6.58	7.63	4.27	4.30	4.82
65	19.10	20.16	21.30	16.65	17.82	15.50	12.00	13.12	14.25	8.65	9.77	10.90	5.22	6.27	7.40	3.65	3.71	4.28

Table 6.3 Wet day correction factors

Number of consecutive days with precipitation	Wet day correction factor
0	1.60
1	0.60
2	0.50
3 and more	0.40

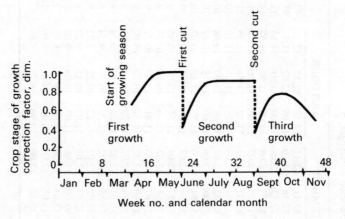

Fig. 6.5 Crop stage of growth correction factor for grass covered soil (after Pierce, 1960).

never reaches the potential rate for subsequent cuts because of physiological reasons. The same sequence of variations is again followed after the second cut but at a still lower level. Finally, as autumn approaches, plant transpiration slowly diminishes. In practice, each segment of the curve may be used independently to accommodate different harvesting dates.

The *correction factor for soil surface cover, K_s,* is particularly relevant for row-crops. The rate of evaporation is assumed to decrease linearly as the percentage surface cover increases to 100 per cent:

$$K_s = 1 - 0.005 \times S_c \qquad \ldots [6.15]$$

Surface cover correction factor, dim $\propto$ Surface cover, %

Including all these correction factors in the evapotranspiration rate formula ensures a close approximation to the actual moisture losses which take place.

6.2.4 Soil moisture model

The sum of runoff, drainage and evapotranspiration gives the total loss of moisture from the soil and, by accounting for the soil moisture gain from precipitation, a soil moisture budget can be maintained on a daily basis. For this purpose, the following data are required:

○ **daily precipitation and sunshine hours;**
○ **mean monthly air temperature;**
○ **latitude;**
○ **soil hydraulic conductivity at field capacity and at saturation;**
○ **soil moisture contents at field capacity, at saturation, and at permanent wilting point;**
○ **soil bulk density.**

All the meteorological data are readily available world wide and the soil data are not difficult to obtain.

The soil moisture model is cumulative in operation and error compensating whenever the soil reaches saturation. Typical results for the annual variation of soil moisture content with rainfall are shown in Fig. 6.6 for a sandy loam soil (Darvel soil series).

6.3 SOIL WORKDAYS

Soil workability is directly related to soil moisture content. The soil moisture model can be used to predict the probability of occurrence of workdays when the soil moisture content is at or below a specific value or *soil workability criterion* over a period of years.

● The **soil workability criterion** is the soil moisture content at or below which the soil is workable.

As soil workability varies from soil to soil, machine to machine and farm manager to farm manager, the adoption of a unique soil moisture value to differentiate between soil workability and non-workability is unrealistic. A procedure has been adopted, therefore, to enable the number of soil workdays to be calculated at

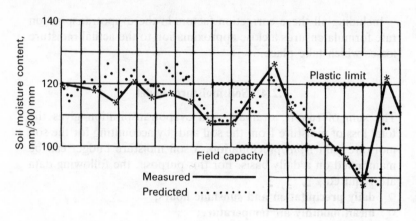

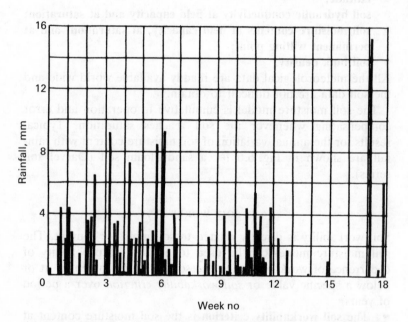

Fig. 6.6 Comparison between measured and predicted soil moisture contents in the plough layer (300 mm) for a sandy loam (Darvel soil series), under grass, in relation to precipitation for the first four months of 1974.

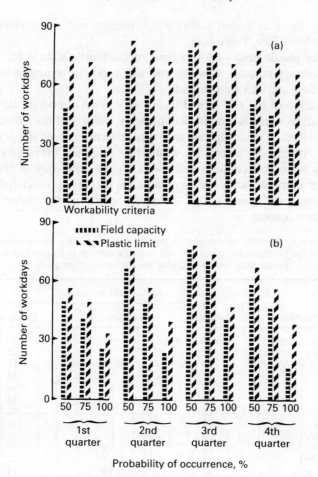

Fig. 6.7 The probability of occurrence of soil workdays for (a) a sandy loam; (b) a clay loam in eastern Scotland.

different levels of soil moisture content or workability criteria. The available days for tillage in each quarter of the year are shown in Fig. 6.7 for two soil types and for two soil workability criteria, namely, the soil moisture contents at field capacity and at the plastic limit.

At a probability level of 100 per cent, that is for 24 years out of 24, work can proceed on a sandy loam on more than two days out of three at a soil workability criterion not exceeding the plastic limit (see Table 6.1).

- The **plastic limit** or lower limit of the plastic state is the soil moisture content at which the soil begins to crumble when rolled out into thin threads.

Cultivation of the soil should ideally be carried out when the soil moisture content is around field capacity and would not normally be undertaken when the soil moisture content exceeds the plastic limit. Attempting to operate within this stricture every year on a clay loam soil confines field work to only one day in three. Rather more workdays are available by taking a greater risk of planning for 12 years out of 24 and by tolerating the higher soil moisture content.

Table 6.4 Monthly field workdays for 18 years out of 24 years in eastern Scotland for three soil types at a soil moisture content not exceeding the plastic limit and at a rainfall not exceeding 10 mm/day

Month	No. of field workdays		
	Light soil	Medium soil	Heavy soil
January	24	18	12
February	23	17	12
March	24	18	15
April	25	19	17
May	26	22	20
June	26	24	24
July	27	26	26
August	26	25	24
September	25	23	20
October	23	19	17
November	22	18	17
December	23	18	14

(*Source*: Eradat Oskoui, 1986)

Monthly field workday data are provided in Table 6.4 for three different soil types and meteorological data for Edinburgh. A field workday is characterised by two factors: firstly, the soil moisture content must not exceed the plastic limit and, secondly, the rainfall must not exceed 10 mm in that day, thereby eliminating unacceptable surface conditions. The effect of other weather constraints, such as the probability of occurrence of snow or frost, is not included in the analysis of soil workdays.

A probability level of 75 per cent, that is 18 years out of 24 years, is used as this is considered appropriate for machinery planning purposes. In the remaining years, the availability of fewer work-days is compensated by extra overtime or by employing more casual labour or contractors to complete the operation.

6.4 SPRAY WORKDAYS

For management planning purposes, the number of spraying occasions available during any period of the year can be determined once the limiting conditions are precisely defined. In addition, however, the operator requires more immediate guidance on the correct tactical decisions of whether to spray or not. Both the analysis of spraying occasions and the development of a spray drift hazard guide are discussed in the following sections.

6.4.1 Spraying occasions

Spraying opportunities at different locations and in different seasons depend not only on the weather but also on an interpretation of meteorological data in relation to the trafficability requirements of different vehicles, the drift hazard with different application techniques and the rainfastness of different spray formulations. There are more spraying occasions for a low ground pressure vehicle than for a conventional tractor. Equally, large spray droplets can be applied in windier conditions than those suitable for a fine mist, and the temperature and precipitation constraints are less stringent for residual sprays than for hormone sprays. The criteria which have been selected to represent the conditions suitable for spraying herbicides are shown in Table 6.5.

The main climatological features of a season cover large geographical areas and, for this reason, the variation across the UK can be indicated by data from five stations:

Kinloss, Grampian – about 50 km east of Inverness;
Aldergrove, N. Ireland – about 20 km west of Belfast;
Waddington, Lincolnshire – about 10 km south of Lincoln;
Boscombe Down, Wiltshire – about 40 km north west of Southampton;
Mount Batten, Devon, Wiltshire – just south of Plymouth.

Table 6.5 Criteria for Spray occasions

Factor	Limiting condition			Residual	Hormone
Operation	Daylight (but between 0600 h and 200 h GMT)				
	Temperature > 1 °C				
	No standing water, glaze, frozen ground or snow lying at 1200 h GMT				
	Visibility period ≥ 5 h consecutive.				
Sprayer	Wind speed ≥ 2 knots (about 1 m/s)				
	Wind speed ≤ 9 knots (about 5 m/s)				
	(measured at 10 m above ground).				
Vehicle	Soil moisture deficit < 5 mm				
	(using a bare soil model and conventional vehicle).				
Chemical				Residual	Hormone
	Temperature	Day maximum > 10 °C			✓
		Next night > 1 °C			✓
	Rel. humidity	< 95%			✓
	Rain	< 1 mm in any hour		✓	
		< 0.1 mm in any hour			✓
		< 0.1 mm in each of 3 hours after spraying			✓
		< 2 mm total in 9 hours commencing 3 hours before spraying			✓

(*Source*: Spackman, 1983)

The average and lower quartile (viz. the value not exceeded on average for one occasion in four) spraying occasions were derived for each month of the year by analysing data for the period from January 1970 to May 1983. The spraying occasions for conventional vehicles are presented in Table 6.6 and those for low ground pressure vehicles in Table 6.7. There is a general trend for the number of spraying occasions to increase from the north and west of the country to the south and east, with higher values inland than near the coast. The trend, however, is not particularly marked, suggesting that the individual stations are influenced almost as much by local surroundings as by geographical position.

The less stringent requirements for applying residual herbicides generally increase the number of spraying occasions, but the impact becomes progressively more marked away from the main growing season.

Table 6.6 Average and lower quartile spraying occasions for conventional vehicles

Chemical	Station	Jan	Feb	Mar	Apr	May	Jun	Jul	Aug	Sep	Oct	Nov	Dec
						Average							
Residual	Kinloss	0	2	7	16	22	24	25	27	14	7	1	0
	Aldergrove	0	0	4	17	21	23	26	27	12	3	0	0
	Waddington	0	0	4	15	21	26	29	27	20	9	1	0
	Boscombe Down	0	0	4	13	19	25	27	28	17	7	2	0
	Mount Batten	0	0	5	15	19	23	25	26	15	7	1	0
Hormone	Kinloss	0	0	1	8	15	18	19	21	12	5	0	0
	Aldergrove	0	0	1	11	16	20	22	23	10	2	0	0
	Waddington	0	0	2	8	16	22	25	21	17	6	0	0
	Boscombe Down	0	0	2	8	15	22	24	24	14	4	0	0
	Mount Batten	0	0	3	11	15	19	21	23	14	6	1	0
						Lower Quartile							
Residual	Kinloss	0	0	1	6	15	22	21	22	5	0	0	0
	Aldergrove	0	0	0	9	16	17	19	20	3	0	0	0
	Waddington	0	0	0	9	16	21	25	22	19	0	0	0
	Boscombe Down	0	0	0	9	12	20	22	24	12	0	0	0
	Mount Batten	0	0	1	9	13	18	20	19	12	0	0	0
Hormone	Kinloss	0	0	0	2	9	16	14	17	3	0	0	0
	Aldergrove	0	0	0	5	10	14	16	17	2	0	0	0
	Waddington	0	0	0	4	12	17	21	17	14	0	0	0
	Boscombe Down	0	0	0	5	8	17	19	17	10	0	0	0
	Mount Batten	0	0	0	6	10	16	16	17	8	0	0	0

(*Source*: Spackman, 1983)

Table 6.7 Average and lower quartile spraying occasions for low ground pressure vehicles

Chemical	Station	Jan	Feb	Mar	Apr	May	Jun	Jul	Aug	Sep	Oct	Nov	Dec
						Average							
Residual	Kinloss	5	9	13	19	23	25	26	30	18	13	7	5
	Aldergrove	6	10	13	23	25	26	29	32	19	16	7	7
	Waddington	5	9	13	20	24	27	31	31	23	17	8	5
	Boscombe Down	6	8	11	16	20	26	28	30	20	14	9	6
	Mount Batten	7	10	12	18	20	23	26	28	21	15	8	7
Hormone	Kinloss	0	1	2	8	15	19	20	23	14	7	1	0
	Aldergrove	0	1	3	14	18	22	24	26	14	10	1	0
	Waddington	0	0	4	10	18	23	26	24	18	9	1	0
	Boscombe Down	1	1	3	9	16	23	24	25	15	8	1	0
	Mount Batten	1	1	5	13	17	19	22	24	17	10	3	1
					Lower Quartile								
Residual	Kinloss	2	6	8	11	15	21	21	25	13	9	5	2
	Aldergrove	4	5	9	18	18	21	21	27	13	13	5	5
	Waddington	4	5	8	16	18	24	26	29	20	14	6	4
	Boscombe Down	4	4	6	11	13	21	22	26	16	11	8	4
	Mount Batten	5	5	8	15	15	19	21	21	16	12	6	4
Hormone	Kinloss	0	0	0	3	9	16	14	20	9	4	0	0
	Aldergrove	0	0	0	7	10	15	16	20	9	7	1	0
	Waddington	0	0	1	4	12	20	22	21	16	7	0	0
	Boscombe Down	0	0	1	5	8	18	17	18	13	5	1	0
	Mount Batten	0	0	1	7	11	16	18	18	13	7	1	0

(*Source:* Spackman, 1983)

For spraying operations, no analysis should ignore reference to soil conditions. In Table 6.5, the vehicle criterion which requires a soil moisture deficit of 5 mm or more for conventional vehicles is somewhat arbitrary since vehicle traction and flotation obviously depends on factors other than soil moisture content (see section 7). In a comparative analysis, however, the choice of the correct parameters (e.g. wind speed, temperature, soil moisture content, etc.) is of major importance, whereas the precise values of the criteria are generally less critical. The average number of spraying occasions for the low ground pressure vehicle (for which *no* soil constraint was imposed) is substantially increased during the winter period, especially for the application of residual herbicides.

When planning spray programmes, it is more appropriate to consider the less favourable years and base machinery management decisions on the lower quartile spraying occasions. Assuming that 10 spraying occasions per month represent the limit below which spraying operations are likely to cause management problems, then the period from October to March inclusive is the most critical. Conventional vehicles are virtually ruled out altogether, and only the least demanding combination of a low ground pressure vehicle applying residual herbicides can achieve a less than optimum chance of successfully completing the spraying operation.

6.4.2 Spray drift guide

The object of the spray drift guide is to demonstrate the effect of different weather and spray patterns on the risk of causing excessive drift, enabling the operator to identify safe periods for spraying which avoid damage to adjacent crops. Spray drift is the proportion of applied chemicals which fails to settle or remain within the target area – chiefly as a result of air movement.

Wind speed, which has the strongest influence on the spraying accuracy of ground-based equipment, is classified by the Beaufort scale, ranging from 0 (calm) to 12 (hurricane force). All wind speed measurements are recorded at a standard height of 10 m above ground level. At a typical boom height of 1 m, the wind strength is about half that at the standard height. During the spraying operation, the chemical fluid has a velocity of about

20 m/s at the jet but this rapidly reduces as the spray fans out towards the ground. As a result, boom height and application technique are critical – the further the jet is from the ground and the finer the spray, the greater the likelihood that small drops will drift away in the wind.

Wind direction is not consistent close to the ground because of surface friction: surface heating also enhances atmospheric turbulence by producing thermals. Sunny days with light winds generate the most unstable conditions, with wind directions being extremely varied. As the earth's surface cools down at night, provided the sky is not overcast, the air in contact with the ground also cools and provides the most stable conditions in the evening and early morning. In order to measure the degree of turbulence, an atmospheric instability scale has been developed with a range from A (unstable) to F (stable).

Other factors play their part in influencing spray patterns. The effect of temperature is clearly seen by the spray droplets evaporating quickly on warm days when the relative humidity is low. As a result, oil-based formulations are used extensively in the tropics to reduce evaporation. The characteristics of the spray also vary with the hydraulic pressure, the orifice size of the jet and the nozzle angle: the higher the pressure and the smaller the hole in the nozzle, the finer the spray; the wider the nozzle angle, the thinner the sheet of liquid and the smaller the average size of droplets.

These four factors – wind speed, atmospheric instability, air temperature and application system – are related by means of a nomograph to give the drift hazard (Fig. 6.8). The hazard rating number – ranging from 1 to 10 – represents the increasing likelihood of drift as the scale rises. Every effort should be made to keep the hazard rating as low as possible, only a very low number being acceptable when using toxic chemicals or when working near susceptible crops.

The dotted example line in Fig. 6.8 shows the sequence which an operator should follow to assess the drift hazard rating. The first task is to measure the wind speed, ideally with a hand-held or mast-mounted anemometer to give an accurate value. Alternatively, a layman's guide to the prevailing wind strength can be obtained from the extent of tree movement (see table with Fig. 6.8). Starting at the appropriate point on the Beaufort Scale

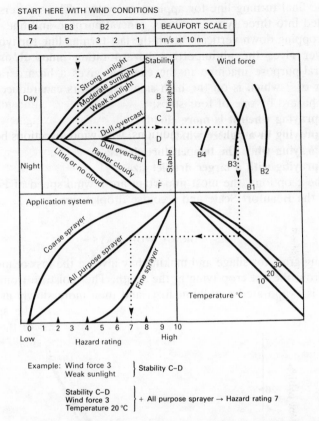

Fig. 6.8 Estimating the drift hazard rating (Anon, *Farmer's Weekly*, 1982).

at the top of the nomograph, draw a vertical line down to the first turning line which represents the prevailing cloud cover for day or night work. The extent of cloud cover in the example is classified as 'weak sunlight'.

The next stage is to move horizontally to the second turning line, the correct curve being identified by the wind speed already used at the starting point. In passing, the dotted example line bisects the atmospheric instability scale. This shows that the conditions are acceptably stable for daytime working.

Draw another vertical line to the third turning line for temperature – in this case 20 °C – and then proceed horizontally to the final turning line for application system. The sprayers are divided into three types – coarse, general-purpose and fine.

Dropping down vertically from the last turning line for type of sprayer gives the all-important hazard rating index. Using a general-purpose machine in the example gives a hazard rating index of 7 which is on the high side. Operators can reduce the drift hazard by one of four steps:

○   **spraying when it is more cloudy;**
○   **spraying in a lighter wind;**
○   **spraying when the temperature drops;**
○   **spraying with a larger droplet size.**

Of these options, the most important are a wind speed of Force 2 on the Beaufort Scale and a coarse droplet size.

6.5   FORAGE WORKDAYS

Wilting grass for silage and making hay involve the loss of moisture from the cut crop lying in the swath. The moisture from the crop is evaporated rapidly at first, and then more slowly as the

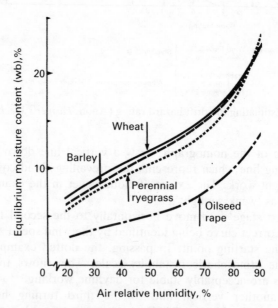

Fig. 6.9   Equilibrium moisture content at 25 °C.

pressure of the water vapour within the herbage approaches the water vapour pressure of the surrounding air. Once the water vapour pressures are equalised, the moisture content of the crop is maintained at an *equilibrium moisture content* which varies with the temperature and relative humidity of the air (Fig. 6.9).

Thus, the drying time can be precisely established from a knowledge of the prevailing atmospheric conditions. Equally, the availability of suitable weather for forage conservation can be identified from the analysis of long term weather records.

The moisture content of agricultural commodities may be calculated on a *wet basis* or on a *dry basis*.

The moisture content of a material on a **wet basis** is the weight of moisture contained in the sample as a percentage of the total weight (including both moisture and dry matter).

The moisture content of a material on a **dry basis** is the weight of moisture contained in the sample as a percentage of its dry matter.

In practical use on the farm, moisture contents are usually quoted on a wet basis:

$$m_{wb} = \frac{W_w}{W_w + W_{dm}} \times 100 \qquad \ldots [6.16]$$

Moisture content (wet basis), % = Wt of water, kg × 100 ÷ [Wt of water, kg + Wt of dry matter, kg]

This produces a non-linear scale, because the values of both the numerator and denominator are changing as the material is dried. To overcome this problem, the dry basis is used in scientific work:

$$m_{db} = \frac{W_w}{W_{dm}} \times 100 \qquad \ldots [6.17]$$

Moisture content (dry basis), % = Wt of water, kg × 100 ÷ Wt of dry matter, kg

Moisture contents on a dry basis are readily converted to a wet basis and vice versa by the formulae:

$$m_{wb} = \frac{m_{db}}{100 + m_{db}} \times 100 \qquad \ldots [6.18]$$

Moisture content (wet basis), % = Moisture content (dry basis), % × 100 ÷ [100 + Moisture content (dry basis), %]

and:

$$m_{db} = \frac{m_{wb}}{100 - m_{wb}} \times 100 \qquad \ldots [6.19]$$

Moisture content (dry basis), % = Moisture content (wet basis), % × 100
÷ [100 − Moisture content (wet basis), %]

Both the dry and wet basis values are used for moisture contents in the subsequent sections.

6.5.1 Field drying in the swath

As the initial moisture content of the freshly cut material defines the start of field drying and the equilibrium moisture content defines the finish of the process, it is possible to identify the swath moisture content whilst drying is in progress from the equation:

$$m_1 = m_e + (m_o - m_e) \times e^{-(kd \times v)} \qquad \ldots [6.20]$$

Swath moisture content, % db = Equilibrium moisture content, % db + (Initial
− Equilibrium Moisture content, % db) × exp (− Drying rate
coefficient × Cumulative vapour pressure deficit, kN h/m²)

The rate of drying is governed by three factors:
○ **maturity of the herbage;**
○ **cumulative vapour pressure deficit;**
○ **drying coefficient.**
Young, leafy material is much wetter relative to the equilibrium moisture content so that it is proportionally easier to lose moisture (although there is a larger volume of water to remove compared with a more mature crop). The cumulative vapour pressure deficit has the greatest influence on overall field drying times.
● The **cumulative vapour pressure deficit** is the difference between the pressure of water vapour at saturation and the actual pressure of the water vapour at the same temperature.
The greater the cumulative vapour pressure deficit, the faster the rate of drying; conversely, during the night, when the cumulative vapour deficit declines because of a heavy dew, the moisture content in the swath rises sharply. Lastly, the value of the drying rate coefficient is affected by the conditioning treatment applied to the swath. As the swath treatment becomes more severe, the

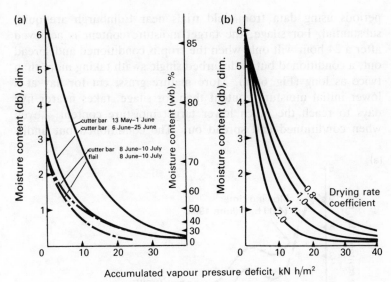

Fig. 6.10 (a) Experimental swath drying curves for different herbage maturity and severity of conditioning treatment; (b) theoretical drying curves for a range of values of the drying rate coefficient (adapted from Spatz *et al.*, 1970; Brown and Charlick, 1972; McGechan, 1985).

drying rate coefficient increases and the rate of drying – as well as the rate of rewetting under adverse conditions – become more rapid.

The interactions of herbage maturity (early versus late cutting dates) and severity of conditioning treatment (cutter bar versus flail mower) are shown in the crop drying curves (Fig. 6.10). The approximate value of the drying rate coefficient for each experimental drying curve can be determined by reference to the most appropriate theoretically derived curves in Fig. 6.10(b).

For practical purposes, it is more important to identify the number of days required to complete the swath drying process. When the experimental drying curves are simulated on a time base, there is not only a general exponential decay as before, but also a cyclic oscillation representing the moisture regained especially during overnight periods of high humidity. Even during unusually good weather conditions in 1984, the simulated drying periods using data from field trials near Edinburgh are quite substantial. For silage, the target moisture content is achieved after a 24-hour wilt only when the crop is conditioned and spread out, a conditioned but undisturbed single swath taking more than twice as long (Fig. 6.11). More mature grass, cut for hay at a lower initial moisture content than for silage, takes nearer five days to reach the much lower target moisture content – even when conditioned and spread out. The conditioned but undis-

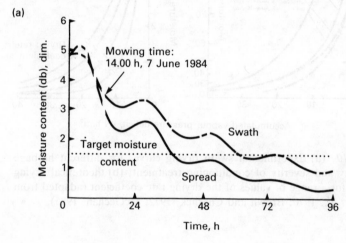

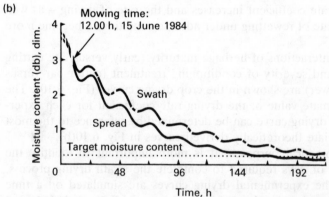

Fig. 6.11 Drying curves for a conditioned single swath for (a) silage and (b) hay, based on experimentally derived coefficients and the 24 hour clock (after Spencer *et al.*, 1985).

turbed single swath takes nine days or more to dry to the target moisture content (Fig. 6.11(b)).

6.5.2 Weather opportunities for forage

Rainfall records over a 25-year period provide an indication of the weather opportunities for forage conservation. Two-day dry spells are required for wilted silage and five-day dry spells for hay (Tables 6.8 and 6.9). In dealing with sequences of different durations, it is more convenient to count the first days of the sequences, since these are the days on which the farmers' decision to cut or not are made. These days are called *starting days*. A dry day is defined as one in which no measurable rainfall is reported (i.e. not more than 0.2 mm in 24 hours). This concept, however, is only a partial guide to the suitability of a particular day for conserving grass. A small amount of rain may be of no practical importance during the warm summer months. On the other hand, a technically 'dry' day may be very humid and have little or no solar radiation, e.g. during coastal fog. To some extent, these two factors should cancel but not counting days with small amounts of rain tends to make the number of longer dry periods unrealistically low. Whilst one light shower in a week of fine weather may rule out several five-day dry spells, the period may have been satisfactory for conventional haymaking.

It is often noted that once two or three successive days have been dry, there is a fairly constant probability of the dry spell extending for a further day. If there are N_2 spells of at least two successive dry days, then the numbers of spells of at least 3, 4, 5, etc. dry days are given by:

$$N_2 \times p; \; N_2 \times p^2; \; N_2 \times p^3,$$

respectively, or more generally by $N_2 \times p^{(n-2)}$ where n is the duration of the required dry spell in days and p is the probability of a recurring weather sequence. In the British Isles, the probability of dry days occurring in sequences, rather than completely at random, ranges from 0.65 to 0.80, and is certainly greater than the ratio of the number of dry days to the total number of days in any specified period. Typically, the number of five-day dry sequences is only one third of the number of two-day dry sequences. Thus, the opportunities for good conservation

Table 6.8 Frequencies of starting days for two-day dry periods at different locations.

Location	May		June		July		August		September	
	1–15	16–31	1–15	16–30	1–15	16–31	1–15	16–31	1–15	16–30
Eastern Britain										
Aberdeen	4.6	6.5	5.5	4.8	4.8	4.8	4.8	4.7	3.3	3.5
Edinburgh	4.6	5.5	5.3	5.8	5.8	6.3	4.3	5.9	3.9	3.4
York	4.7	6.9	7.8	6.0	6.8	7.0	5.0	7.0	5.7	5.4
Cambridge	6.4	7.4	8.5	7.6	7.5	8.0	6.6	8.2	7.7	6.0
Western Britain										
Ayr	4.8	6.4	5.4	5.1	5.9	5.5	4.8	6.0	3.5	3.8
Penrith	4.5	6.5	5.9	5.1	5.2	5.2	4.6	5.9	4.1	4.0
Aberystwyth	4.5	6.8	6.6	5.3	5.6	7.2	4.8	6.5	4.9	4.3
Shrewsbury	5.1	7.1	6.7	6.6	7.6	7.4	5.2	7.2	6.6	5.4
Plymouth	4.9	7.2	8.4	6.9	7.1	8.3	5.6	7.9	6.2	5.2

Table 6.9 Frequencies of starting days for five-day dry periods at different locations

Location	No. of starting days for five-day dry sequences									
	May		June		July		August		September	
	1–15	16–31	1–15	16–30	1–15	16–31	1–15	16–31	1–15	16–30
Eastern Britain										
Aberdeen	1.8	2.9	2.0	1.4	1.6	1.7	2.0	1.8	0.8	1.1
Edinburgh	1.4	2.7	2.1	2.8	1.9	2.5	1.3	2.6	1.5	1.1
York	1.7	3.9	3.3	3.1	3.3	2.4	2.4	3.3	2.1	2.1
Cambridge	3.1	4.2	4.8	4.2	4.5	3.8	3.6	4.7	3.8	2.9
Western Britain										
Ayr	2.2	3.0	2.4	2.3	2.3	1.9	1.6	3.1	1.0	1.4
Penrith	1.9	3.3	2.3	2.2	2.2	2.3	1.9	3.0	1.3	2.1
Aberystwyth	2.3	3.8	3.3	2.4	1.9	2.8	2.9	3.5	1.8	1.9
Shrewsbury	1.4	3.6	3.6	3.8	3.8	3.4	2.2	3.4	2.8	2.6
Plymouth	2.8	3.4	5.2	4.5	3.7	4.7	3.1	4.2	3.2	2.0

decrease rapidly with systems requiring longer dry day sequences, and this only serves to re-emphasise the unreliability of haymaking compared with ensiling as a forage conservation process.

In a maritime climate, early cutting improves the chances of better weather and coincides with the time when grass is at its most nutritious. Dry day sequences are at a maximum during the second half of May and the first half of June, and so is the solar radiation. The increased cloudiness gives less sunshine after the summer solstice than before. For second-cut silage, the most favourable period occurs in the second half of July. Then for a late cut the second half of August is not to be despised, provided that the shortening days and the increased likelihood of heavy dews are borne in mind.

6.6 COMBINE HARVESTING WORKDAYS

During the growth of the cereal crop, sap moisture is present and grain moisture content may be very high. When no more dry matter is produced, the crop is said to have reached *morphological maturity*. The moisture content of wheat at this point in time approaches 50 per cent, wet basis, and thereafter falls steadily at about 2 per cent, wet basis, per day (Fig. 6.12). Once the moisture content is reduced to about 30 per cent, wet basis in the UK, it becomes influenced by ambient conditions in a way similar to that already described for forage (see section 6.5) and the equilibrium moisture content varies with the temperature and relative humidity of the air (Fig. 6.13).

After the crop has reached morphological maturity, the start of harvesting depends on the method of harvesting. The appropriate 'ripeness' or *technological maturity* of the crop is the stage of maturity at which no subsequent reduction in dry matter yield or quality will occur naturally during the harvesting, drying or storage processes. In the days of the binder, harvesting would commence at moisture contents of about 40 per cent, wet basis, whereas now the majority of farmers wait for the grain moisture content to fall below 21 per cent, wet basis, for combine harvesting.

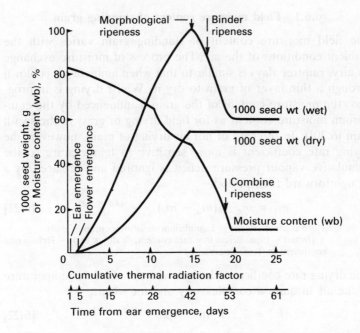

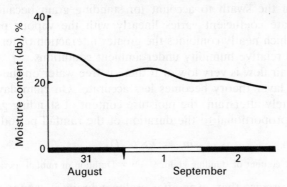

Fig. 6.12 The development of wheat in relation to a cumulative thermal radiation factor, $\Sigma t\sqrt{R}$, where t is the average daily temperature, °C, and R is the total daily radiation, cal (after Philips and O'Callaghan, 1974).

Fig. 6.13 The moisture content of a standing crop of Midas barley, showing both a diurnal variation and a gradual moisture reduction (after Smith *et al.*, 1981).

6.6.1   Field moisture content of standing grain

The field moisture content of standing grain varies with the ambient conditions of the air. The process of moisture exchange on airy, rainfree days is similar to that when ambient air is blown through a thin layer of grain to dry it. Whilst drying is in progress, the moisture content of the grain is influenced by the equilibrium moisture content as for field drying of grass in the swath (eqn [6.20]). In the case of hot air drying of grain, however, the drying rate coefficient is more sensitive to temperature so the cumulative vapour pressure deficit is ignored and replaced by a straightforward time period:

$$m_1 = m_e + (m_o - m_e) \times e^{-(kd \times \Delta t)} \quad \dots \text{[6.21]}$$

Grain moisture content, % db = Equilibrium moisture content, % db
    + [Initial − Equilibrium moisture content, % db] × exp (− Drying rate coefficient × Time interval, h)

The drying rate coefficient changes with the absolute temperature of the air in quite a complex way and for wheat:

$$kd = 7.2 \times 10^{-6} \times e^{-5044/(T_a + 273)} \quad \dots \text{[6.22]}$$

Drying rate coefficient ∝ Exponential function of Air temperature, °C

which, at a temperature of 15 °C, gives a value for the drying rate coefficient of 0.15. Whilst high temperature, thin layer drying theory can explain field drying of grain at more modest temperatures, it seems more appropriate to adapt the equation for drying forage in the swath to account for standing grain because the drying rate coefficient varies linearly with the vapour presure deficit which neatly combines the greater interaction of temperature and relative humidity under ambient conditions.

If the air flow is very low or if there is free water on the grain, the thin layer theory becomes less accurate. On calm days and immediately after rain, the moisture content of standing grain is directly proportional to the duration of the rainfall period:

$$m_g = 2.8 \times D_r \quad \dots \text{[6.23]}$$

Moisture content of standing grain, % db ∝ Duration of rainfall, period, h

with the proviso that, even after prolonged rain, it cannot exceed an upper moisture content limit of 50 per cent, dry basis.

Typical observed values of the field moisture content for barley show a gradual downward trend over a three-day period (Fig. 6.14). It is also evident that there is a marked diurnal variation, with the lowest moisture content in the late afternoon or early evening. This fluctuation lags behind, by about four or five hours, the daily changes in the relative humidity of the air which has a minimum relative humidity at approximately mid-day.

6.6.2 Criteria for combine harvesting

The simplest criterion for the available combine harvesting time is based upon rain-free days. This, however, takes no account of either the effect of rainfall intensity on the grain moisture content during the days preceding the dry day or of the influence of moisture fluctuations in the standing crop on machine performance. Although grain moisture content is important because it affects the amount of drying required, the effect of rainfall is less immediate on the grain than on the straw. As the surface of the straw becomes damp, the coefficient of straw to metal friction increases and restricts the throughput of the combine harvester. Once the surface of the straw is really wet, the free water acts as a lubricant, reducing the friction again. Whilst the effect of rain or dew on the straw to metal friction is a transitional one, the efficiency of both the primary and the secondary grain/straw separation mechanisms in the combine harvester is reduced in the presence of water. Consequently, combine harvesting usually ceases after a significant fall of rain, taken as being in excess of 1.27 mm.

The direct result of the rainfall limit is that combine harvesting can take place below that limit, with an important proviso to include the effect of prolonged wet periods. Instead of rain-free days, combine harvesting can start only on days when the *discounted sum* of the past rainfall is less than 1.27 mm.

- The **discounted sum** is the rainfall in the past 24 hours plus 20 per cent of the previous day's discounted sum.

This is the same as taking a geometrically decaying sum of past rainfall, with each day given 20 per cent of the weighting of the succeeding day. In a more recent study in Eastern Scotland, the discounted sum of rainfall was replaced by a single, but slightly higher, rainfall limit. A statistical comparison between weather records and observations of work days from a telemetry survey of combine harvester operations established the criterion such that:

combine harvesting can take place when the rainfall in the previous 24 hours is less than 1.4 mm.

The extent of the harvesting season depends on the starting and finishing conditions. A standing crop of barley which has a grain moisture content of 30 per cent, wet basis, on 1 August requires a further 10 days to reach combine harvesting ripeness at a grain moisture content of 21 per cent, wet basis. Any crop still remaining uncut 60 days later, that is by 9 October, may be regarded as a total loss. As the season progresses, the time available for harvesting on a good day decreases. The potential harvesting day length of 9 hours declines at a rate of 0.02 h/day, so that only 7.6 hours are available on a good day at the end of the harvesting season. On this basis, the time available for combine harvesting, averaged over a 10-year period, is in the region of 350 hours for the main cereal growing areas of England.

6.7 SUMMARY

Soil workability is governed by the soil moisture content and the weather pattern.

The soil moisture budget includes the effect of rainfall, runoff, drainage and evapotranspiration.

Surface runoff depends on the moisture retention characteristics of the soil.

Drainage flow is obtained from the hydraulic conductivity of the soil at field capacity and at saturation.

Evapotranspiration rate is the rate of evaporation from a free water surface, corrected for changes in the rate of transpiration of any crop cover with availability of water in the root zone and with stage of plant development.

From the daily estimation of soil moisture content, the number of soil workdays when the soil moisture content is below a given value can be determined for any locality and soil type.

Spraying occasions are based on an interpretation of meteorological data in relation to the trafficability requirements of different vehicles, the drift hazard with different application techniques and the rainfastness of different spray formulations.

Drift hazard is presented as a nomograph which combines the effect of four factors – wind speed, atmospheric instability, air temperature and application system.

The rate of crop drying in the swath is affected by the maturity of the herbage, the cumulative vapour pressure deficit and the conditioning treatment.

Weather opportunities for forage conservation are given as sequences of dry days, two dry days being required for wilting silage and five dry days for haymaking.

Once standing grain has reached the correct stage of maturity, combine harvesting can take place when the rainfall in the previous 24 hours is less than 1.4 mm, the time available being in the region of 350 hours.

7

TRACTIVE PERFORMANCE AND POWER SELECTION

OUTLINE

Soil strength; Cohesive and frictional soils; Cone penetration resistance; Clay ratio; Tractive thrust; Rolling resistance; Wheelslip and travel reduction; Tyre mobility; Coefficient of traction; Tractive efficiency; Cross-ply and radial-ply tyres; Specific ploughing resistance; Ploughing performance predictor; Implement hitching and tractor weight distribution; Weight transfer, weight addition and ballast; Tractor drawbar performance predictor; Engine-limited and slip-limited power output; Least cost ploughing; Tractor fleet selection for sequential operations.

APOSTROPHE — FETCHING HOME THE MILLSTONE

'Among the "parts, pertinents, and privileges" granted under a baronial charter in the feudal times, was the mill, all the lands of the barony being astricted or "thirled" thereto, forming the mill "sucken". The tenants were bound to have their corn ground at the mill to which they were thirled, even if it was not the nearest to them. Each person in the sucken had to pay mill multures, and to perform certain services, such as assisting to bring home a new millstone when required.

In the time when there were neither properly made roads nor wheeled vehicles capable of bearing so heavy a load, the process of fetching home the millstone from a distance of perhaps a dozen miles or more was an onerous business. The mode adopted was simply that of trundling it on its edge all the way, by the most direct route available. They got a long and stout stick, which was called the "spar", put through the eye of the millstone, and firmly wedged there. The spar projected from two to three feet on one side, and perhaps fifteen feet, or more, on the other, the long lever being used to keep the stone on its edge, the other in the way of guidance as the stone moved onward. Over the millstone was fixed a rough wooden frame. Four, or perhaps six, horses were yoked to the front of this frame, which had a steering tree attached behind, while its construction admitted of the spar turning round like the axle of the "tumbling cart". One experienced man steered; another kept by the short end of the spar; while the general body of the suckeners managed the long end, or held on behind by ropes attached to the frame, to prevent the millstone running off on the downward gradients.

Despite every precaution, the millstone would occasionally get too much way on a declivity and overpower all concerned, creating dire confusion; or by some unhappy chance it would have its equilibrium so disturbed as to get suddenly upset on the short end of the spar, throwing the hapless suckeners, who hung grimly on at the other end, hither and thither, or tilting them up in the air. These experiences were neither pleasant nor safe!'

(Source: Anon, 1877)

7.1 INTRODUCTION

Tractor power is utilised in two ways by transmitting the engine power both through the driving wheels as traction to provide the drawbar power required for draught implements, and through the power take-off shaft – as well as the hydraulic system – to provide mobile support for attached machines. Tractive performance is essential for heavy draught operations, such as ploughing, and this drawbar power dictates the total tractor power requirement on an arable farm. On livestock farms, tractor selection is based on the need to transfer a greater proportion of the engine power directly to drive forage harvesting machinery, but tractive performance cannot be completely ignored in any field activity.

In both traction and cultivation, the forces created by the passage of the machine cause the soil to deform and rupture. As the stress–strain relations involved in soil deformation are extremely complex, an empirical approach has been developed in which soil strength is measured by means of the resistance to the penetration of a standard 'walking stick' called a cone penetrometer. Using the cone penetration resistance as an indication of soil strength, it is possible to establish the mobility of vehicles and the draught of various soil-engaging implements operating on different soil types and under different weather conditions.

Even using simplified procedures, the number of machine, soil and weather factors is still considerable. For practical purposes, the various factors are combined into two series of linked graphs, the '*Ploughing performance predictor*' and the '*Tractor drawbar performance predictor*' to provide a visual guide for the selection of the most suitable tackle under a given set of circumstances. These technical analyses are then combined with an economic appraisal to select the optimum tractor/machine combinations for least-cost ploughing and to assess the tractor fleet size for three sequential operations to establish winter cereals.

7.2 SOIL STRENGTH

The strength of the soil can be tested in the field with a walking stick: the more compact the soil, the greater the force required to push in the stick. Soil moisture content also makes a difference, but to a much greater extent in a heavy soil than in a sandy soil. The shank which glides into a wet sticky clay hardly dents the surface when the soil is dry.

A more objective measure of the penetration resistance of the plough layer can be obtained from an instrument called a cone penetrometer (Fig. 7.1). This is simply a standardised 'walking stick' which has a conical tip and a meter to indicate the pressure, or cone penetration resistance, developed as the penetrometer is forced into the ground. The cone penetration resistance has been widely used to predict the tractive capabilities of off-road vehicles, the draught forces in tillage operations and the compaction related to vehicle traffic. Although the cone penetration resistance of the soil is readily obtained by direct measurement in the field, it is more appropriate in the simulation of traction and draught to be able to evaluate the cone penetration resistance from the basic soil and weather variables.

7.2.1 Cohesive and frictional soils

When a block of soil is displaced relative to another, it shears along a failure plane and the force required to move the soil

Fig. 7.1 The cone penetrometer being used to evaluate soil strength in the plough layer.

within the failure zone depends on the shear strength of the soil. This shear strength is derived from a combination of two soil properties, the *cohesive strength* and the *frictional resistance*.

- **Cohesion** is the molecular attraction of like substances in contact.

A saturated clay extracted from the natural deposits in a pottery claypit is highly cohesive and, as its strength is not altered when extra pressure is applied to the failure plane, the frictional resistance is very low (Fig. 7.2). In contrast, dry sand on the beach

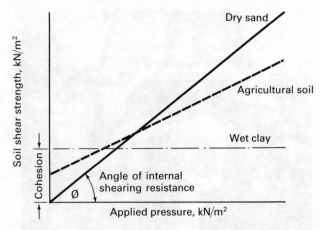

Fig. 7.2 The effect of the cohesive and frictional properties of different soil types on the soil shear strength.

derives its strength almost entirely from its frictional resistance which is determined by its angle of internal shearing resistance (or 'friction angle').

- The ratio of the frictional resistance to the applied pressure is equivalent to the *tangent* of the **angle of internal shearing resistance.**

The angle of internal shearing resistance increases as the particle size becomes coarser. For practical purposes, the upper value of 45 degrees for this angle is roughly equivalent to the angle of repose of a conical heap of dry sand. Thus, the frictional resistance is unlikely to exceed the applied pressure because the tangent of this limiting angle of friction is unity.

The majority of agricultural soils are neither of these extreme types and, instead, range from heavy clays with predominantly cohesive properties, through medium loams with both cohesive and frictional properties, to sandy soils with mainly frictional characteristics. Despite the wide range of soil types, the shearing force is described by a single relationship which is of fundamental importance to the study of soil mechanics:

$$H = c \times A + W \times \tan \phi \qquad \dots [7.1]$$

Shearing force, kN = Cohesion, kN/m^2 × Area in contact, m^2 + Applied load, kN × Tangent of Angle of internal shearing resistance, deg.

This relationship clearly shows that the strength of cohesive soils depends on the area in contact, whereas the frictional soils rely on loading imposed on the failure surface.

The forces generated by any soil failure process, either in tillage or in traction or by cone penetration, depend on the volume of soil displaced. As the variation in the compressibility of the topsoil affects the extent of the failure zone, the area of the failure surface is not readily determined. This compressibility, in turn, means that the contribution towards the applied load from the mass of the soil, within the undefined failure zone, cannot be specified either. For, this reason, it is necessary to adopt, within a theoretical framework, an empirical approach for the understanding of traction available from, and the tillage draught required for, different soils.

7.2.2 Cone penetration resistance

When a cone penetrometer is pushed into the ground, it forces the soil laterally from its path along a logarithmic spiral shaped boundary to form a pressure bulb (see Fig. 7.4(b)). The size of this pressure bulb is governed by the angle of internal shearing resistance: the larger the angle, the greater the mass of soil affected. Consequently, the penetration resistance of a frictional soil varies according to the degree of compaction as indicated by the specific weight of the soil. In more cohesive soils, the smaller friction angle results in a smaller size of the pressure bulb but the reduction in frictional resistance may be more than offset by the effect of the cohesive resistance. This gives a general expression for the cone penetration resistance in the form:

$$CI = [K_c \times c + K_\phi \times \gamma \times \tan \phi]e^{(\pi \times \tan \phi)} \qquad \ldots [7.2]$$

Cone penetration resistance, MPa = [Cohesive coeff., dim. × Cohesion, MPa + Frictional coeff., m × Soil specific weight, kN/m^3 × Tangent of Friction angle, deg.] × Logarithmic spiral factor, dim.

For a standard size of cone, the value of the cohesive coefficient in eqn [7.2] is approximately constant because the effect of the surface area of the pressure bulb is taken care of by the logarithmic spiral factor. A constant value of the frictional coefficient additionally requires a standard penetration depth – taken as the median depth of the plough layer – because the frictional resistance is dependent on the volume rather than on the surface area of the pressure bulb. Even though unique values for the coefficients can be established experimentally, the remaining properties of the individual soils vary enormously and measurement is very tedious. Both the angle of internal shearing resistance and the cohesion of the soil are substantially affected by soil type. In addition, the moisture variation in agricultural soils is very large and has a major influence on the cohesive strength of the soil.

Soil types are classified by the mechanical analysis of particle size according to the proportions of clay, silt and sand. These three fractions – four with a further division into fine and coarse sand – can be combined into a single number called the clay ratio. As the clay fraction has cohesive properties by virtue of its chemical bonds, the ratio of clay to silt and sand, R_c, indicates the relative importance of the cohesion to friction. A soil containing more clay is characterised by a higher clay ratio and is a more cohesive material at any given moisture content. As the soil becomes wetter, however, the cohesive strength falls rapidly, declining virtually to zero above the liquid limit when heavy soils turn into fluid mud. In other words, the cohesive component of the penetration resistance is directly proportional to the clay ratio and inversely proportional to the moisture content.

Sandy soils, containing a smaller proportion of clay than heavy soils, have a lower clay ratio. The tangent of the angle of internal shearing resistance is, therefore, also linked to the clay ratio, such that:

Table 7.1 Relation between the tangent of the angle of internal shearing resistance and values of the clay ratio.

Angle of internal shearing resistance (Ø), deg	Tan $Ø = 1/(1+2\ R_c)$	Clay ratio (R_c)
45	1	0
40	0.84	0.10
35	0.70	0.21
30	0.58	0.35
25	0.47	0.57
20	0.36	0.87
15	0.27	1.34
10	0.18	2.35
5	0.09	5.20
1	0.02	28.00

$$\tan \phi = 1/(1 + 2 \times R_c) \qquad \ldots [7.3]$$

Tangent of Angle of internal shearing resistance, deg. α 1/Clay ratio, dim.

For a dry sand with a clay ratio of zero, this gives an angle of internal shearing resistance of 45 degrees, whilst, at the other end of the scale, a very heavy soil with over 80 per cent clay content results in a representative friction angle of only 5 degrees (Table 7.1).

Including the effects of both moisture content and soil type, the complete expression for the cone penetration resistance at median plough depth becomes:

$$CI = [K_c \times R_c \times e^{-0.1 \times \theta/(1+R_c)} + K_\phi \times \gamma/(1 + 2 \times R_c)] \times$$
$$e^{\pi/(1 + 2 \times R_c)} \qquad \ldots [7.4]$$

Cone penetration resistance, MPa = [Cohesive coeff., dim. × Clay ratio, dim.
   × Exponential function of Moisture content, % w/w, and Clay ratio, dim.
   + Frictional coeff., m × Soil specific weight, kN/m³/Function of Clay
   ratio, dim.] × Exponential function of Clay ratio, dim.

From experimental studies, the values of the cohesive and frictional coefficients are 3.62 and 6.63×10^{-3}, respectively. Over a range of clay ratios, the relative contributions of the frictional and cohesive components of the cone penetration resistance equation are shown in Fig. 7.3 for three soil moisture contents and one soil specific weight. The curve for the frictional component decays quickly to a modest value with increasing clay ratio. The curves for the cohesive components all emanate from the origin and rapidly establish different levels of direct proportionality with the clay ratio, the drier soils contributing the highest cohesive strength component.

Typical values of the clay ratio for different soil types are given in Table 7.2. The variation of the cone penetration resistance with soil moisture content, soil specific weight and soil type is shown in Fig. 7.4(a). In a purely frictional dry sand, the soil moisture content has no effect on the cone penetration resistance but the variation in soil specific weight has a considerable influence, as represented by the width of the horizontal band. As the clay ratio increases for heavier soil types, the effect of the soil specific weight diminishes but the changes in soil moisture content become increasingly more important – lower soil strength at high

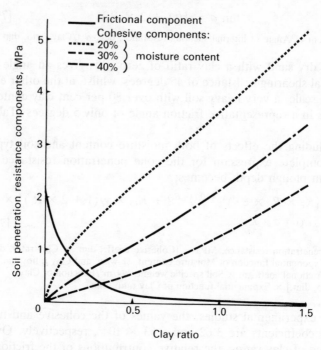

Fig. 7.3 The effect of clay ratio on the frictional and cohesive components of the soil penetration resistance at various moisture contents and a soil specific weight of 14 kN/m².

Table 7.2 Typical values of the clay ratio for different soil types

Soil type	Particle size fractions, %				Clay ratio (R_c)
	Course sand	Fine sand	Silt	Clay	
Coarse sandy soil	72	19	3	2	0.02
Sandy soil	45	32	10	8	0.08
Light loam	41	29	14	9	0.10
Medium loam	24	21	25	20	0.25
Heavy loam	25	9	24	29	0.40
Clay soil	18	9	21	36	0.55
Silt soil	1	17	59	13	0.15

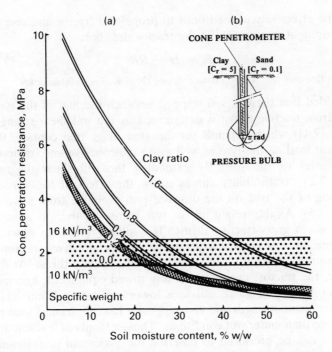

Fig. 7.4 (a) The effect of soil moisture content and clay ratio on the cone penetration resistance for a band of soil specific weights from 10–16 kN/m³; (b) the relative size of the pressure bulb formed at the base of a cone penetrometer for clay (LHS) and sand (RHS).

moisture content and very much higher cone penetration resistances at low moisture contents compared with the cone penetration resistance in the purely frictional sand.

7.3 TRACTIVE PERFORMANCE

Tractor engine torque is converted into tractive thrust at the ground drive wheels by pushing against the soil in contact with the lugs of the tyres – the weaker the soil, the lower the tractive thrust available. Not only is there less traction; there is also more resistance to the forward movement of the wheel as it sinks more deeply into soft ground. This restricts the net force or pull which can be effectively utilised both to propel the tractor and also to pull draught implements at the tractor drawbar:

$$F_h = H - RR \qquad \qquad \ldots [7.5]$$

Drawbar pull, kN = Tractive thrust, kN − Rolling resistance, kN

Provided that there is soil trapped between the lugs of the tyre, the gross tractive thrust is determined by the soil shear strength (eqn [7.1]) which depends on the area of ground contact, the vertical load on the tyre as well as on the cohesive and frictional properties of the soil. By extending the soil shear diagram (Fig. 7.2), 'trafficability curves' show the effect of the vertical loading of the tyre on the drawbar pull available from a wheel (Fig. 7.5). As the weight on the tyre increases, there is a linear increase in gross tractive thrust. The rolling resistance which is minimal for low vertical loads on the tyre progressively becomes more substantial until it finally exceeds the tractive thrust. At this point, the tractor, even without any towed equipment, becomes immobilised. The same soil at a lower moisture content has a greater shear strength and can support a heavier vertical load on the tyre than under wet conditions. Thus, a family of trafficability curves can be produced to describe the maximum performance

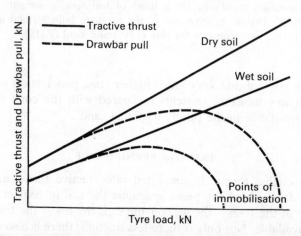

Fig. 7.5 For different soil conditions, trafficability curves show the drawbar pull available from a drive wheel after deducting the rolling resistance from the gross tractive thrust.

of the driving wheels under different conditions and on different soils.

These trafficability curves indicate the *maximum* pull which can be developed by the tyre under the particular soil conditions. In practice, this maximum tractive performance may not be required and only a proportion of the shear strength of the soil under the tyre is mobilised. When less resistance is demanded from the soil, less soil deformation occurs under the wheel and the *wheelslip* is reduced.

- **Wheelslip** is the reduction in distance travelled per revolution of a wheel in work compared with the distance travelled by the same wheel developing zero thrust on the same operating surface.

A driven wheel always has some wheelslip, albeit small, just to provide the necessary horizontal force for propelling itself forward; equally, a towed wheel always has some skid – negative wheelslip – to overcome the frictional resistance of the wheel rotating on the axle. In order to be strictly correct, therefore, the distance travelled by a wheel developing zero thrust is the mean of the distances travelled by both a self-propelled wheel (developing no drawbar pull) and a towed wheel (overcoming no braking force) on the same operating surface and during the same number of wheel revolutions.

$$s = 1 - d_1/d_0 \qquad \dots [7.6]$$

Wheelslip, dim. = 1 − Distance travelled per rev. in work, m ÷ Distance travelled per rev. at zero thrust, m

The inaccessibility of the zero wheelslip point is clearly shown by reference to the wheel force-torque-slip diagram (Fig. 7.6) for a self-propelled wheel and a driven wheel (both generating some wheelslip), as opposed to a towed wheel and a braked wheel (both causing some wheel skid). In each case, the vertical load on the wheel, W, is the same – and this includes both the weight of the wheel itself and any vertical loading imposed by the vehicle on which the wheel is mounted. The soil reaction force, G, comprises two components, a vertical component which must be equal and opposite to the vertical load on the wheel, and a horizontal component which must be equal and opposite to the net effect of any tractive thrust and rolling resistance. This horizontal force is given by the drawbar pull for a driven wheel

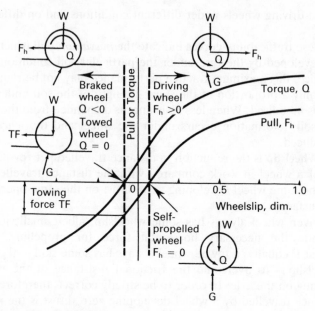

Fig. 7.6 The pull/torque/wheelslip relation for wheels on soil (adapted from Brixius and Wismer, 1978).

and by the towing force for a towed wheel. Any wheel torque, Q, is counterbalanced by the moment of the ground reaction force which is offset from the centre of the axle. For this reason, the ground reaction force acts through the centre of the axle only for a towed wheel because the wheel torque is zero. For a driven wheel, the torque is being transmitted from the engine, whereas the torque acting in the opposite direction on the braked wheel is generated by the wheel being used to power a ground driven mechanism or by applying brakes to reduce the travel speed.

When an implement is towed behind a tractor, the additional drawbar pull increases the requirement for tractive thrust from the drive wheels which try to meet the extra demand at the expense of a higher level of wheelslip. The tractive performance of a tractor is evaluated by means of a pull/slip test (Fig. 7.7). As the drawbar pull is gradually increased, the incremental changes in wheelslip are small initially, but become steadily larger towards the maximum drawbar pull until the point of

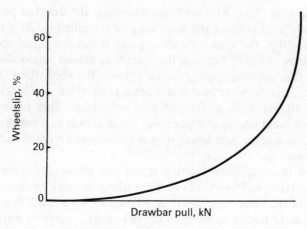

Fig. 7.7 The tractive performance of a tractor is indicated by the level of wheelslip at different drawbar pulls, the maximum pull occurring at complete wheelspin.

immobilisation is reached at 100 per cent wheelslip. Although the steepening of the curve is often used to suggest the optimum wheelslip for peak tractive performance, an accurate result is only possible by taking into account the travel speed as well as the drawbar pull to give the effect of drawbar power on the

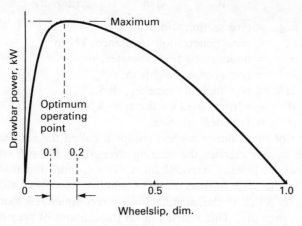

Fig. 7.8 The optimum wheelslip for maximum drawbar power.

wheelslip (Fig. 7.8). At a very low wheelslip, the drawbar power is virtually zero because the drawbar pull is negligible. At a very high wheelslip, the drawbar pull is greatest but the travel speed is now close to zero because the tractor is almost immobilised. Again, the drawbar power is very low. Between these two extremes, the drawbar power reaches a peak value which occurs around 10 per cent to 20 per cent wheelslip. This optimum operating point occurs at a wheelslip which is only just noticeable in the field and is much lower than is commonly accepted in field operations.

Whilst the maximum tractive thrust can be readily obtained from the basic soil properties and the wheel or track parameters, the use of the cone penetration resistance allows a much broader evaluation of tractor mobility in the field over a range of soil and loading conditions.

7.3.1 Tyre mobility

Using dimensional analysis, the main traction parameters – including soil strength, applied load and tyre configuration – can be grouped into dimensionless ratios and linked together to form the *tyre mobility number*:

$$MN = \frac{CI \times b \times d}{W} \times \sqrt{\frac{\delta}{h}} \times \frac{1}{1 + b/(2 \times d)} \quad \ldots [7.7]$$

where: b = tyre section width, m;
   CI = cone penetration resistance, kN/m^2;
   d = undeflected tyre diameter, m;
   h = tyre section height, m;
   MN = tyre mobility number, dim.;
   W = vertical load on the tyre, kN;
and δ = tyre deflection, m.

The first of these dimensionless groups is called the *soil number* because it characterises the bearing strength of the soil relative to the wheel loading imposed on it. The second dimensionless group is the *tyre stiffness number* and its maximum value is limited to $\sqrt{0.2}$ at the manufacturer's recommended load and inflation pressure. This restriction on the amount of tyre deflection, measured statically on a hard surface, of one fifth of the

Radial ply

Cross ply

Fig. 7.9 ´Ply construction of tyres (adapted from Inns and Kilgour, 1978).

section height of the tyre is to avoid side wall damage. The last group in eqn [7.7] is the *tyre shape number* to account for the variation in the geometry of the more circular cross section of cross-ply tyres and the flatter, more rectangular cross section of radial-ply tyres (Fig. 7.9). Although the tyre mobility number in itself appears to have little practical significance, various aspects of tyre performance are related to it, such as the *coefficient of traction* and the *coefficient of rolling resistance*.

The coefficient of traction provides a measure of the pull from a driven wheel in relation to the vertical loading imposed on it.

● The **coefficient of traction** is the ratio of the drawbar pull to the vertical load on the tyre.

The highest value of this coefficient on a sandy soil – neglecting any traction losses – is equivalent to the tangent of the angle of internal shearing resistance (see eqn [7.1]) and, therefore, is unlikely to exceed unity. During field work, the normal range is from 0.6 to 0.8. As the shear strength of the soil is gradually mobilised with increasing deformation or slip, so the drawbar pull builds up to a maximum value for the particular ground conditions. This, in turn, means that the value of the coefficient of traction, C_T, builds up to a maximum value, $(C_T)_{max}$ (Fig. 7.10). Compared with a track on a firm soil, tyres require a greater amount of slip to develop maximum performance. As the operating surface becomes softer, the rolling resistance is increased and it is not until higher values of wheelslip that any drawbar pull becomes available. Even then, the coefficient of traction only builds up slowly and attains an unacceptably low maximum value.

The relationship between the coefficient of traction and the wheelslip is represented by an expression of the form:

$$C_T = (C_T)_{max} \times (1 - e^{-k \times m}) \qquad \ldots [7.8]$$

Coefficient of traction, dim. = Maximum coefficient of traction, dim.
  × (1 − Exponential function of Rate constant, dim. and Wheelslip, dim.)

In order to calculate the coefficient of traction at any given wheelslip, it is necessary only to designate the value of the maximum coefficient of traction which is:

$$(C_T)_{max} = \frac{(F_h)_{max}}{W} = 0.796 - \frac{0.92}{MN} \qquad \ldots [7.9]$$

Maximum coefficient of traction, dim. = Maximum drawbar pull, kN ÷ Vertical load on tyre, kN
  = Function of Tyre mobility number, dim.

and the way in which this coefficient increases with wheelslip is indicated by the rate constant. The slope of coefficient of traction versus slip curve at the origin is given by the product of the rate constant and the maximum coefficient of traction:

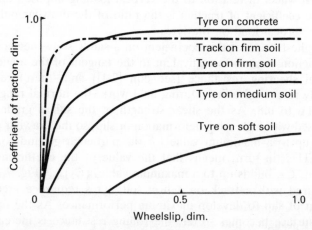

Fig. 7.10 The variation of the coefficient of traction with wheelslip on different surfaces for a tyre compared with that for a track on firm soil (after Brixius and Wismer, 1978).

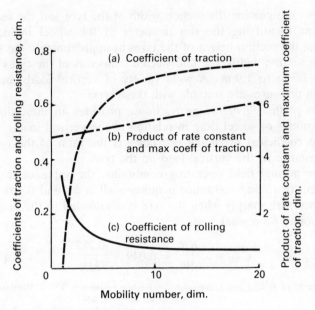

Fig. 7.11 The variation of the main traction parameters with tyre mobility number: (a) maximum coefficient of traction; (b) product of rate constant and the maximum coefficient of traction; (c) coefficient of rolling resistance (adapted from Gee-Clough, 1980).

$$k \times (C_T)_{max} = 4.838 + 0.061 \times MN \quad \ldots [7.10]$$

Rate constant, dim. × Maximum coeff. of traction, dim.
= Function of Tyre mobility number, dim.

The variation in both the maximum coefficient of traction and the rate constant with the tyre mobility number is shown in Fig. 7.11. The rapid initial rise in value of the maximum coefficient of traction with increasing tyre mobility number gradually slows as the coefficient of traction approaches the highest value of rather less than 0.8 in very firm ground conditions, whereas the product of the rate constant and the maximum coefficient of traction has a modest linear increase. These empirical expressions are valid for stubble, ploughed and cultivated fields only; loose surface trash or grass cover on hard, dry soil gives results which were substantially different from all the others. The expressions are also valid for tyre sizes which range from 12.4–36 to 18.4–38, the first number designating the section width of the tyre and the second number identifying the rim diameter of the wheel in inches. Taking the section height of the tyres as approximately the same as the section width, the undeflected diameters of the tyres vary from 1.5 m to 1.9 m. A wide variety of vertical loads and inflation pressures are suitable with these tyres.

The *coefficient of rolling resistance* provides an indication of the amount of wheel drag in relation to the vertical load.
- The **coefficient of rolling resistance** is the ratio of the rolling resistance to the vertical load on the tyre.

Under normal field operating conditions, the value of the coefficient of rolling resistance is quite small at around 0.1 to 0.3, but increases sharply when the tyre is overloaded to the point of immobility on a weak soil:

$$C_{RR} = \frac{RR}{W} = 0.049 + \frac{0.287}{MN} \qquad \ldots [7.11]$$

Coefficient of rolling resistance, dim. = Rolling resistance, kN ÷ Vertical load on the tyre, kN.
= Function of Tyre mobility number, dim.

The trend for the effect of tyre mobility number on the coefficient of rolling resistance is almost a mirror image of the coefficient of traction (Fig. 7.11). The rolling resistance declines sharply once the ground is hard enough to support the wheel with little sinkage.

7.3.2 Tractive efficiency

Tractive performance is assessed from the pull, torque and slip characteristics of the tyre. These factors are combined to give the *tractive efficiency*.
- The **tractive efficiency** is the ratio of the tractive power output to the axle power input.

It is of major importance to know the pull and slip at which maximum tractive efficiency is achieved because they determine the size of the implement and the travel speed required to make full use of the available power.

The tractive power output of the wheel is given by the product of the drawbar pull and the travel speed, accounting for slip:

$$P_o = F_h \times v_o \times (1 - s) \qquad \ldots [7.12]$$

Tractive power output, kW = Drawbar pull, kN × Theoretical forward velocity at zero slip, m/s × (1 − Wheelslip, dim.)

The axle power input from the transmission is determined by the torque, Q, required at the driving wheel and the rotational velocity of the axle, ω:

$$P_i = Q \times \omega = Q \times v_o/r_r \qquad \ldots [7.13]$$

Axle power input, kW = Axle torque, kN m × Rotational velocity, rad/s
= Axle torque, kN m × Theoretical forward
velocity, m/s ÷ Rolling radius of tyre, m

This torque requirement is equivalent to the moment of the total tractive thrust being developed at the interface between the tyre and the soil:

$$Q = (F_h + RR) \times r_r \qquad \ldots [7.14]$$

Axle torque, kN m = (Drawbar pull, kN + Rolling resistance, kN)
× Rolling radius of tyre, m

As the tractive efficiency, η_t, is a ratio, the forces in eqns [7.12] to [7.14] may be replaced by the coefficients of traction and rolling resistance, and the rolling radius of the tyre is eliminated, so that:

$$\eta_t = \frac{C_T \times (1 - s)}{C_T + C_{RR}} \qquad \ldots [7.15]$$

Tractive efficiency, dim. = Coefficient of traction, dim. × (1 − Wheelslip, dim.)
÷ (Coefficient of traction, dim. + Coefficient of rolling resistance, dim.)

The tractive efficiency varies with wheelslip in a similar way to that for drawbar power (Fig. 7.8), but the important difference is that the optimum operating point at maximum tractive efficiency occurs at a similar level of wheelslip, regardless of the ground conditions normally encountered in practice. This optimum operating point can be determined mathematically by differentiating eqn [7.15] with respect to wheelslip and obtaining the maximum point where the gradient of the tractive efficiency curve, $d\eta_t/ds$, is zero:

$$(1 - s) \times \frac{dC_T}{ds} = \frac{C_T \times (C_T + C_{RR})}{C_{RR}} \qquad \ldots [7.16]$$

Gradient of curve = Function of Coefficient of traction, dim., Coefficient of
rolling resistance, dim. and Wheelslip, dim.

The coefficient of rolling resistance and the coefficient of traction (through its maximum value and the rate constant) are all functions of the tyre mobility number. Inevitably, therefore, the wheelslip at maximum tractive efficiency is also a function of mobility number:

$$s_{opt} = 9 + 19/MN \qquad \ldots [7.17]$$

Optimum wheelslip, % = Function of Tyre mobility number, dim.

A good average figure for the wheelslip at maximum tractive efficiency is 10 per cent because, fortunately, it varies very little with mobility number under normal conditions (Fig. 7.12).

Finally, now that the optimum wheelslip has been established in terms of the mobility number, the maximum tractive efficiency can be similarly derived by using the appropriate values in eqn [7.15], to give the expression:

$$(\eta_t)_{max} = 78 - 55/MN \qquad \ldots [7.18]$$

Max. tractive efficiency, % = Function of Mobility number, dim.

The effect of mobility number on the maximum tractive efficiency is also shown in Fig. 7.12.

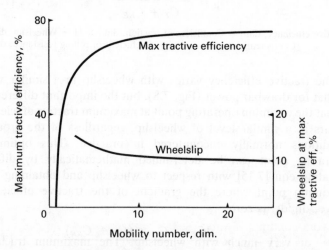

Fig. 7.12 The variation of wheelslip at maximum tractive efficiency and maximum tractive efficiency itself with tyre mobility number.

As well as providing a means of optimising tractor performance (see Section 7.5), the results of these analytical studies demonstrate that a very large volume of information on the performance of numerous different sizes and types of tyres can be presented in a concise form. This information is displayed in the *Handbook of Agricultural Tyre Performance*.

7.3.3   Tyre performance data

The tyre performance data are estimated for four representative field conditions: good, average, poor, and bad. The 'good' condition is equivalent to hard dry grassland, with a cone penetration resistance of 2 MPa and no visible tyre sinkage. The 'average' condition represents a typical stubble in a good state for ploughing, with a cone penetration resistance of 400 kPa and only shallow ruts. For the 'poor' condition found when ploughing wet stubble or during secondary cultivations, fairly deep ruts and some traction difficulties are to be expected on ground with a cone penetration resistance of 250 kPa. Lastly, the 'bad' condition represents the worst conditions for heavy draught work which, though inadvisable, is unavoidable. Typically difficult operations are ploughing land after root harvesting in a very wet autumn or preparing a seedbed in a wet spring when the cone penetration resistance is only 150 kPa.

Five parameters are used in the Handbook to define the performance of a tyre:

coefficient of traction at 20 per cent wheelslip;
coefficient of rolling resistance;
maximum tractive efficiency;
coefficient of traction at maximum tractive efficiency;
and wheelslip at maximum tractive efficiency.

The various terms are defined in the previous section, and an example of the tabulated information for one tyre load at a range of tyre inflation pressures is given in Table 7.3.

The coefficient of traction is given at 20 per cent wheelslip because this is usually close to the highest value at which a tractor can work continuously at reasonable efficiency. Rolling resistance does not vary much with wheelslip. The maximum tractive efficiency is probably the most useful value because it determines the drawbar power which can be transmitted by the tyres for a

Table 7.3 A sample page from the 'Handbook of Agricultural Tyre Performance'

1600 kg (3530 lb) Load
6 Ply Tyres

Tyre size		12.4/11–36		13.6/12–38		16.9/14–30	
Inflation, bar		1.5		1.1		0.8	
pressure, lbf/in^2		22		16		12	
Field conditions	kN	lbf	kN	lbf	kN	lbf
Pull at	Good	11.8	2650	11.8	2650	11.8	2650
20% slip	Average	7.3	1640	7.5	1690	7.6	1710
	Poor	6.4	1440	6.7	1510	6.8	1530
	Bad	4.7	1060	5.3	1190	5.5	1240
Rolling	Good	1.2	270	1.1	250	1.1	250
resistance	Average	1.7	380	1.6	360	1.6	360
	Poor	2.1	470	1.9	430	1.9	430
	Bad	2.8	630	2.6	580	2.4	540
Maximum	Good		76		77		77
tractive	Average		67		68		69
efficiency, %	Poor		61		63		64
	Bad		50		54		55
Pull at	Good	6.3	1420	6.4	1440	6.4	1440
maximum	Average	5.8	1300	5.9	1330	5.9	1330
efficiency	Poor	5.4	1210	5.5	1240	5.6	1260
	Bad	4.7	1060	4.9	1100	5.0	1120
Slip at	Good		10		10		10
maximum	Average		13		13		12
efficiency, %	Poor		15		15		14
	Bad		20		18		18

(*Source*: Dwyer, *et al.*, 1975)

given input power at the axle. A knowledge of the coefficient of traction and wheelslip at maximum tractive efficiency is essential to ensure that both fall within the range of practical operation, and to enable designers to calculate the optimum wheel loads, net drawbar pulls and tractor gear ratios for a given field.

For all cross-ply tyres with an open centre tread pattern to aid self cleaning (Fig. 7.13), source of manufacture has little influ-

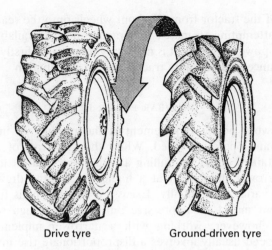

Drive tyre   Ground-driven tyre

Fig. 7.13 Drive tyres have an open tread pattern to aid self cleaning, the lug angle being reversed for ground driven tyres (after Inns and Kilgour, 1978).

ence on tractive performance, but there are some specific aspects of tyre design and use which are relevant. Increasing the lug height beyond the minimum necessary to penetrate surface vegetation or a slippery top layer has a detrimental effect on tractive performance. Radial-ply tyres provide 5 to 8 per cent extra pull compared with cross-ply tyres. Running in the furrow increases the pull at 20 per cent wheelslip by up to 28 per cent because the tyre obtains a better grip, but the rolling resistance also increases by up to 140 per cent. Running in the rut formed by a preceding tyre increases the pull at 20 per cent wheelslip by an average of 7 per cent and decreases the rolling resistance by an average of 11 per cent.

In order to achieve the highest possible rate of work at different operating speeds, the choice of the most appropriate tyre size for a tractor of a certain engine power is largely governed by the vertical load on the driving wheels. This vertical load comprises not just a fixed proportion of the static weight of the tractor but also ballast from wheel weights or water-filled tyres. Furthermore, the angled draught force to pull an attached implement imposes an extra downward force on the rear wheels of the tractor as well as causing a transfer of some of the static weight of the tractor from the front wheels onto the rear wheels. Before attempting to match the tractive power available to the drawbar power demand, it is necessary to examine firstly drawbar performance and secondly tractor weight distribution.

7.4 DRAWBAR POWER DEMAND

The drawbar power requirement is the product of implement draught and operating speed. Whilst the same rate of work can be maintained either by pulling a wide implement at low speed or a narrower implement at a higher speed, the drawbar pull demand varies substantially. Excessive drawbar pulls from wide equipment may look impressive but can cause high wheelslip. Equally, however, working with a narrower implement at a higher speed usually involves a disproportionate rise in draught. Between these two extremes, there is an optimum operating point which is dependent on the sensitivity of the implement draught to changes in the operating speed.

7.4.1 Plough draught

Ploughing is the most demanding routine draught operation on an arable farm, the need for subsoiling being less frequent. As a means of eliminating plough size from a performance comparison, the draught of a plough is divided by the cross sectional area of the furrow(s) and quoted as a specific ploughing resistance. The specific resistance of a mouldboard plough is a combination of the quasi-static soil shearing resistance and a dynamic component increasing with the square of the velocity and influenced by the lateral direction angle of the mouldboard tail. Using the cone penetration resistance as the measure of soil strength, the specific resistance of the plough, Z, is:

$$Z = 37.6 \times CI + 23.9 \times \gamma \times v^2 \times (1 - \cos \lambda)/g \quad \ldots [7.19]$$

Specific ploughing resistance, kN/m^2 α Cone penetration resistance, MPa + Soil specific weight, kN/m^3 $\times$ (Velocity, m/s)^2 $\times$ Function of Mouldboard tail angle, deg. $\div$ Gravitational acceleration, m/s^2

Taking into account the variation in soil type, soil moisture content and compaction, the soil penetration resistance determines a basic level of plough draught which can vary enormously from rock hard dry clay to a friable sandy loam. Superimposed on this basic draught force is the combined effect of plough body shape and operating speed.

A long, highly twisted, helicoidal mouldboard with a small tail angle (Fig. 7.14) can be used effectively over a wide speed range, whereas the draught of a shorter, more cylindrical body increases more rapidly with speed. As the rear end of the mouldboard is responsible for the final turning of the furrow, a large tail angle throws the soil further and the power wastage increases with the *square* of the operating speed. This dynamic effect also involves the weight of the furrow slice being moved. While the loose soil left after harvesting potatoes is easy to turn over, the heavily trafficked ground on the headlands develops a much greater resistance. The consolidation of the topsoil under stubble lies somewhere in between these two limits.

Along with the details of soil strength, operating speed and plough type, the total plough draught is governed by the furrow width, the depth of ploughing and the number of plough bodies in use. The operating speed is again required for the determination of the drawbar power and, by identifying a suitable tractive efficiency (somewhat arbitrarily at this stage), it is then possible to obtain an estimate of the tractor engine power using a series of linked graphs in the from of a *Ploughing Performance Predictor* (Fig. 7.15).

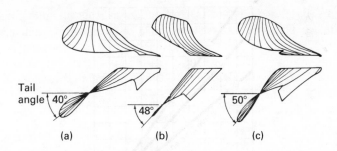

Fig. 7.14 Mouldboard shapes showing the vertical and horizontal lines as well as the tail angle: (a) the highly twisted helicoidal body for heavy land; (b) the cylindrical body for maximum disintegration and deeper work; (c) the semi-helicoidal body form to meet intermediate requirements (after Söhne, 1959).

7.4.2 Ploughing performance predictor

The *Ploughing Performance Predictor* can be used in several ways. It shows the rate of work obtainable. It also demonstrates the sensitivity of the power requirement to the various characteristics of a particular plough. Then again, if the tractor is already available, a check can be made on the number of bodies, the depth of work or even the optimum travel speed.

Example 1: Rate of work

The work rate in the field depends on the travel speed, so start there (Fig. 7.16). As it is assumed that the turning time is kept to a minimum, the field efficiency is fixed at 80 per cent (see section 3). The two remaining factors to be selected are the furrow width and the number of plough bodies. By changing direction at each of these decision points, the user moves diagonally across the chart on a zigzag course which leads to the rate of work scale.

Many farmers are more familiar with the number of weeks

Fig. 7.15 The 'Ploughing Performance Predictor' chart (from Terratec, 1982).

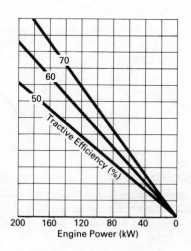

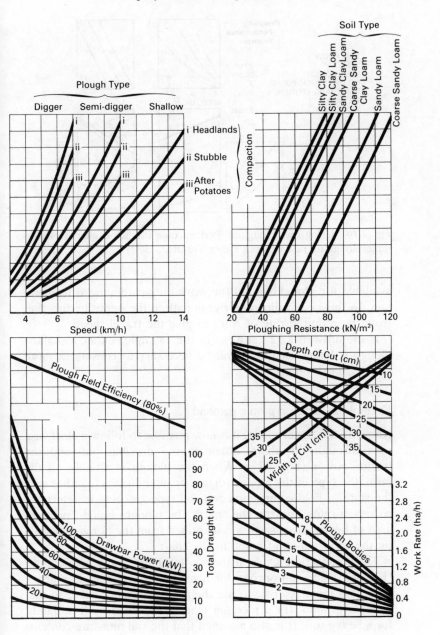

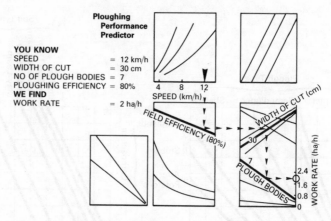

Fig. 7.16　Using the 'Ploughing Performance Predictor' chart to determine ploughing rate of work (from Terratec, 1982).

available for ploughing and the work output they need to maintain. In this case, follow the zigzag path in the opposite direction from the required rate of work to give the travel speed.

If the answer is 12 km/h, as in the example, either a shallow plough is being used or it will be necessary to consider a more realistic work output.

Example 2: Tractor power demand

Knowing the travel speed, it is now possible to follow a clockwise path through the chart, obtaining the drawbar pull and the tractor power (Fig. 7.17). When the plough is already on the farm, the type of the plough (digger or semi-digger) together with the state of field compaction and soil type are used to select the correct turning lines on the chart. The plough draught per unit area of furrow cross section is identified from the horizontal scale on the top right-hand graph of the chart.

The compaction levels in the top left-hand graph of the chart (Fig. 7.15) are appropriate for a medium loam soil. For heavy soils, it is more realistic to use the adjacent higher curve so that 'stubble ground' for a medium loam is replaced by 'headlands' for a heavy soil. It is also assumed that the soil moisture contents built into the top right-hand graph of the chart approximate to

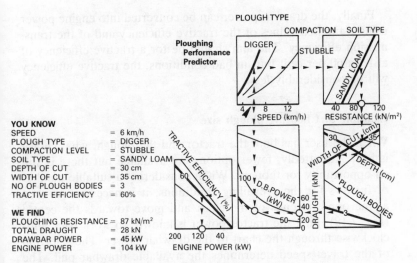

Fig. 7.17 Using the 'Ploughing Performance Predictor' chart to evaluate tractor power demand (from Terratec, 1982).

the field capacity values. If a farmer is prepared to work in wetter conditions, he can look at the effect of this option by selecting the adjacent heavier 'soil type' line as a turning value. It must be remembered, however, that working in wetter conditions is a double-edged sword. Plough draught is reduced but only at the expense of poorer tractive efficiency which must be adjusted to a lower value in the bottom left-hand graph of the chart.

Converting the plough draught per unit of furrow cross-section to drawbar pull only requires a knowledge of the width of the furrow, the depth of working and the number of plough bodies in use. There is quite a lot of scope for choice here. A shallower depth and fewer bodies make a big difference to drawbar pull. Such choices are used in the field far too seldom.

The product of drawbar pull and travel speed gives the drawbar power. On the chart, the drawbar power is readily obtained at the intersection of the vertical travel speed line and the horizontal drawbar pull line. From the intersection point, move parallel to the nearest constant power curve until reaching the vertical scale at the edge of the drawbar power graph (bottom centre, Fig. 7.17).

Finally, the drawbar power can be converted into engine power using estimated values of the tractive efficiency and of the transmission efficiency. The aim should be for a tractive efficiency of at least 60 per cent but, in bad conditions, the tractive efficiency will be considerably less.

Example 3: Choosing plough size

When the user has both the tractor and the plough, what choice is left? Admittedly, fewer options are available but there is still an opportunity for thought. When considering a suitable number of furrows for the prevailing conditions, it is necessary to start at the opposite ends of the chart and move towards the centre (Fig. 7.18). First the tractor power is known, so work back anticlockwise through the chart to the drawbar power. The selection of the travel speed determines the available drawbar pull. The path through the chart cuts horizontally across the 'number of furrows' graph (bottom right) but the correct number of furrows is not known at this stage. Transfer to the second starting point at the travel speed and, working clockwise, finish with a vertical

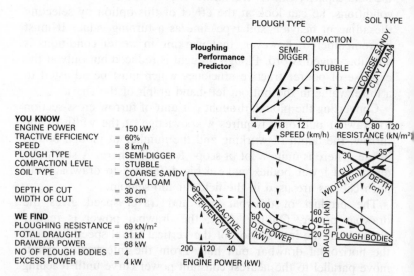

Fig. 7.18 Using the 'Ploughing Performance Predictor' chart to choose the size of plough (from Terratec, 1982).

line in the 'number of furrows' graph. The nearest number of whole furrows to the intersection point provides the answer which, in the example, is four furrows.

Other factors such as best plough type and travel speed and best depth of working can be selected by trial and error. It is not difficult to master the use of the chart and certainly much easier to examine the various possibilities on paper than it is in the field. Apart from setting up the plough, how many changes are normally made to the ploughing procedures once the tractor leaves the steading?

7.5 TRACTOR PERFORMANCE

Pulling any load, attached to the tractor drawbar or to the three point linkage, generates a draught force which tends to tilt the tractor/implement combination like a see-saw pivoting on the ground contact patch of the rear wheels. This action increases the weight being carried by the rear wheels and can be used to advantage to increase the traction available from a rear wheel driven tractor, just when it is most required. This redistribution of weight from the front axle of the tractor to the rear is less useful on a four-wheel driven tractor because the tractive capability of the front wheels is then under-utilised. For optimum tractive performance, therefore, four-wheel driven tractors are designed to be 'nose heavy'.

Whilst tyre performance data are obtained from single wheel tests with known *static* weights, their application to the evaluation of tractor performance requires an understanding of the *dynamic* weight distribution between the front and rear axles when the tractor is in work.

The dynamic weight distribution of the tractor is influenced by the method of attaching the implement or machine to the tractor. The three methods of implement attachment are:

○ **full mounted (integral);**
○ **semi-mounted (semi-integral);**
○ **trailed.**

An implement which is *fully mounted* on the three-point linkage of a tractor must form a stable combination both in the transport position and in work. If the weight of the equipment is too heavy, the front axle of the tractor will lift off the ground with the loss of steering control. There is, however, no danger of the tractor rearing over backwards because of the three-link connection with the implement. Within limits, extra weights can be added to the front of the tractor to counterbalance the weight of the equipment mounted on the rear.

When a very long implement is attached to the tractor linkage, either its weight exceeds the lifting capacity of the tractor hydraulic system or the centre of gravity of the implement is so far back from the tractor that the implement stays on the ground and the front of the tractor lifts off the ground instead. As neither result is acceptable, it is necessary to resort to *semi-mounted attachment*. The front of the implement is still attached to the three-point linkage of the tractor – or sometimes just to the two lower links – but some of the weight of the implement is also carried on a tail wheel.

Trailed equipment is only connected to the tractor at the drawbar hitch point and has the least effect on the dynamic weight distribution of the tractor, as explained in the following analysis.

7.5.1 Tractor weight distribution

The forces acting on an implement in work are:

○ **implement weight:**
○ **soil forces,** e.g. forces for cutting and turning the soil when ploughing;
○ **support forces,** e.g. forces from a depth wheel or skid;
○ **hitch forces** transmitted to the mounted implement through the three ball-ends of the linkage.

For simplicity, the effect of the soil and support forces are represented by a single horizontal draught force, acting at the soil surface (Fig. 7.19). The combination of the implement weight and the horizontal draught force gives the resultant draught force, F_r, which acts at an angle, θ, to the horizontal. When the tractor is working at uniform speed, this resultant draught force must be exactly balanced by the resultant of the hitch forces.

Thus, the dynamic weight distribution of the tractor is determined by the magnitude and direction of the resultant draught force and the static weight of the tractor acting through its centre of gravity. The component of the static weight of the implement

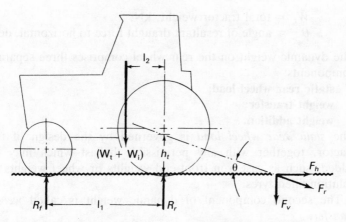

Fig. 7.19 Some of the static weight of the tractor/implement combination, $(W_t + W_i)$, is transferred from the front wheels onto the rear wheels through the action of the resultant draught force F_r acting at an angle θ to the horizontal, whilst weight addition may also occur from the vertical component of the resultant draught force acting on the rear wheels.

carried on the tractor can be included in the static rear wheel load, provided that it is excluded from the resultant draught force which, in turn, alters the angle of its line of action. Summation of the vertical and horizontal components of these forces and moments results in the following expression for the dynamic weight on the rear wheels of the tractor:

$$R_r = (W_t + W_i) \times (l_1 - l_2)/l_1 + F_h \times h_t/l_1 + F_h \times tan\ \theta$$
$$\dots [7.20]$$

Rear wheel reaction, kN = Static rear wheel load, kN + Weight transfer, kN
        + Weight addition, kN

where: h_t = vertical distance from resultant draught force to
            rear wheel contact patch, m;
    F_h = drawbar pull, kN;
    l_1 = tractor wheelbase, m;
    l_2 = horizontal distance from tractor/implement cen-
            troid to rear axle, m;
    R_r = rear wheel reaction, kN;
    W_i = static weight of implement carried on the tractor,
            kN;

W_t = total tractor weight, kN;

θ = angle of resultant draught force to horizontal, deg.

The dynamic weight on the rear wheel comprises three separate components:

○ **static rear wheel load;**
○ **weight transfer;**
○ **weight addition.**

The *static rear wheel load* is governed by the design of the tractor, together with any permissible ballast which may be added, such as cast iron front-end weights or wheel weights or solution-filled tyres.

The second component of dynamic weight is called *weight transfer*.

● **Weight transfer** is that part of the tractor weight which in the static condition rests on the front wheels, but which is transferred to the rear wheels through the turning moment of the resultant draught force.

A greater value of the horizontal component of the resultant draught force and a higher line of action of the resultant draught force above the rear wheel contact patch combine to increase the weight transferred to the rear wheels. The line of action of the resultant draught force can be altered by adjusting the geometry of the three-point linkage and by varying the 'draught' control setting of the hydraulic system to influence implement response to soil penetration (see section 8.7.2).

The third component of dynamic weight is *weight addition* (not to be confused with ballast).

● **Weight addition** results from the vertical component of the resultant draught force increasing the weight on the rear wheels.

If the resultant draught force is horizontal, there is no weight addition, for example when towing a balanced, four-wheel trailer which is supporting its all-up-weight through its own axle. There is still some weight transfer, unless the drawbar hitch is lowered to ground level. This is an important safety feature in drawbar design: attempting an excessive drawbar pull causes the front of the tractor to rear up and lowers the drawbar hitch point to the ground, eliminating any further increase in overturning moment. The tractor may become immobilised by wheelspin, but it does not overturn.

Compared with the static rear axle load of the tractor listed in the manufacturer's specification, the combination of the static weight of the implement and dynamic weight transfer typically adds about 65 per cent with fully mounted equipment, 45 per cent with semi-mounted equipment, and 25˙ per cent with trailed equipment. These proportions are not constant and are a function of the actual implement used (light or heavy, close coupled or very long), and of the soil variation which alters both the horizontal draught force and the vertical penetration force.

7.5.2 Tractor drawbar performance predictor

Once the tractive efficiency curve has been obtained for a tyre over a range of travel reduction values on a particular soil (see Fig. 7.10), the drawbar performance of any two-wheel driven tractor can be determined graphically by means of the *Tractor Drawbar Performance Predictor Chart* (Fig. 7.20). Tractor drawbar performance is derived from two ratios, the coefficient of traction and the tractive efficiency (see section 7.3).

The coefficient of traction or pull to weight ratio can be given as:

$$C_T = F_h/(W_{rs} + F_h \times C_{dwt}) \qquad \ldots [7.21]$$

Coefficient of traction, dim. = Drawbar pull, kN ÷ (Static tractor weight on rear wheels, kN + Drawbar pull, kN × Coefficient of dynamic weight transfer, dim.)

Instead of considering the total dynamic weight acting on the rear wheels as a single entity, it is divided into two parts, one being the static weight, W_{rs}, and the other being the dynamic weight component governed by the drawbar pull, F_h, and the coefficient of dynamic weight transfer, C_{dwt}. This partition enables the effect of the different implement attachment methods to be illustrated on the chart by varying the value of the coefficient of dynamic weight transfer.

The second ratio incorporated in the chart is the tractive efficiency, or the ratio of drawbar power output from the rear wheels to axle power input:

$$\eta_t = F_h \times v_{sp} \times (1 - s_{sp})/P_i \qquad \ldots [7.22]$$

Tractive efficiency, dim. = Drawbar pull, kN × Forward velocity at zero pull, m/s × (1 − Travel reduction, dim.)/Axle power input, kW

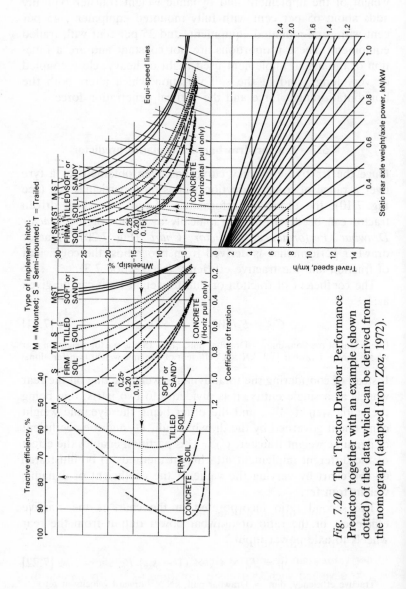

Fig. 7.20 The 'Tractor Drawbar Performance Predictor' together with an example (shown dotted) of the data which can be derived from the nomograph (adapted from Zoz, 1972).

This relationship for tractive efficiency follows American practice by including the travel reduction which is zero at the self-propulsion point with no drawbar pull, instead of the wheelslip which is zero at the point of no thrust (see Fig. 7.6).

- **Travel reduction** is the reduction in distance travelled per revolution of a wheel in work compared with the distance travelled by the same wheel developing zero pull on the same operating surface.

The actual travel speed is then determined from the travel reduction with reference to the forward velocity at zero pull, or 'no load' speed.

The basis for constructing the *Tractor Drawbar Performance Predictor Chart* is obtained by equating these two expressions for the drawbar pull and rearranging the terms to group the tyre performance criteria together, such that:

$$W_{rs} \times v_{sp}/P_i = \eta_t \times [(1/C_T) - C_{dwt}]/(1 - s_{sp}) \qquad \ldots [7.23]$$

where: C_{dwt} = coefficient of dynamic weight transfer, dim.;
$\quad\quad\quad C_T$ = coefficient of traction, dim.;
$\quad\quad\quad P_i$ = axle power input, kW;
$\quad\quad\quad s_{sp}$ = travel reduction. dim.;
$\quad\quad\quad v_{sp}$ = forward velocity at zero pull, m/s;
$\quad\quad\quad W_{rs}$ = static tractor weight on rear wheels, kN.
$\quad\quad\quad \eta_t$ = tractive efficiency, dim.

The chart contains average tyre performance curves for four surface conditions:

concrete	– typical for standard tractor tests;
firm soil	– relatively high strength with low sinkage, typical of a grass ley;
tilled soil	– loose soil after ploughing and one discing;
soft or sandy	– very low strength with high sinkage.

The curves within each surface condition represent typical values for the coefficient of dynamic weight transfer for different implement attachment methods:

0.65 for fully mounted equipment;
0.45 for semi-mounted equipment;
0.25 for trailed equipment.

Other values can be interpolated if they are available. The chart is also designed such that easily obtained or readily measurable inputs can be used: for instance, drawbar pulls are based upon static tractor weight on the rear wheels rather than the more elusive dynamic weight for a tractor/implement combination in work; and travel speeds are entered at the 'no-load' speed rather than the actual speed which is itself a function of wheelslip.

Example 1: Checking drawbar performance

The chart can be entered at any point where values are known or can be assumed. The normal point of entry is at the tractor 'no load' travel speed. With the chart, the expected travel reduction, drawbar pull, travel speed and drawbar power can be determined. A typical path is shown on the chart.

The first step is to identify the power available at the rear axle. Whilst this figure is not generally included in a tractor specification, a reasonable estimate is obtained by assuming that the ratio of the power take-off power to rear axle power is 0.96 for gear type transmissions. In the example, therefore, the tractor developing 70 kW at the power take-off has a maximum power at the rear axle of 67 kW. It is also known that the static tractor weight on the rear wheels is 40.2 kN and that the 'no load' travel speed is 7.5 km/h, operating on firm soil with a fully mounted implement. Thus, the ratio of static rear weight to axle power is 0.6 kN/kW.

Selecting a travel speed of 7.5 km/h, move horizontally to the first turning point on the 0.6 kN/kW line. Proceed vertically upwards to meet the 'fully mounted' line for a firm soil.

From this second turning point, there are alternative paths to follow. The actual travel speed is identified by returning down the chart along the curved equi-speed lines which compensate for travel reduction and thereafter back to the travel speed axis via the same 0.6 kN/kW turning line. This yields a value of 6.5 km/h for the actual travel speed. Proceeding horizontally from the second turning point produces three further intersections. Firstly, the travel reduction is shown to be 13.5 per cent which is consistent with the actual travel speed of 6.5 km/h.

The second family of curves in the centre of the chart represent coefficient of traction against travel reduction. Continuing through these curves to the 'fully mounted' line for firm soil, the third turning point leads down to the coefficient of traction, in this case equal to 0.73. Thus, the drawbar pull is 0.73 × 40.2 kN, that is 29.3 kN.

Passing straight through the third turning point, the horizontal path is extended to meet the family of tractive efficiency curves for the different surface conditions. The tractive efficiency is directly above this final turning point on the firm soil curve. In the example, the tractive efficiency is 0.78, so that the drawbar power becomes 0.78 × 67 kW, namely 52 kW.

Example 2: Tractor ballasting

The *Tractor Drawbar Performance Predictor Chart* can also be used to determine the approximate ballast required for tractors in the field. From the left-hand side of the chart, it can be seen that the maximum tractive efficiency occurs at a travel reduction of 5 per cent on concrete, 10 per cent on firm soil, 13 per cent on tilled soil and 15 per cent on soft or sandy soil. It is desirable to operate close to the maximum efficiency point in order to achieve as much drawbar power as possible. In practice, it is preferable to ballast for a travel reduction which is greater than that required for maximum drawbar power because drawbar power drops more sharply for low travel reductions than for high (through excessive rolling resistance) and because the travel reduction decreases with part load operations (tractors being unable to operate at above maximum power).

By retracing the path through the chart, the rear wheel weight required for peak tractive efficiency is obtained from the lower right-hand graph. Ballast calculations may result in a vertical loading which is beyond the capacity of the tyres through using lower than recommended travel speeds. As the travel speed is increased, the weight on the rear wheels can be reduced because less drawbar pull is required to utilise the drawbar power.

7.5.3. Engine-limited and slip-limited power output

The drawbar performance of a tractor may be restricted by two factors:

○ **maximum engine power;**
○ **wheelslip.**

The output power is limited by the engine power available in the *higher* gears, whereas it is limited by wheelslip in the *lower* gears. The lower of the alternative values is the actual maximum power available in any particular gear. Whilst the graphical procedure described in the previous section helps to visualise the important factors involved in maximising drawbar performance, an analytical approach is more versatile in establishing the overall performance of a wider range of rear-wheel driven and four-wheel driven tractors equipped with fully mounted ploughs.

Engine-limited power output

The calculation of engine-limited power is very straightforward. The engine torque at maximum power is multiplied by the transmission efficiency and the appropriate gear ratio to give the torque available at the driving wheels. This torque is divided by the rolling radius of the wheels and the total rolling resistance is subtracted to identify the drawbar pull, F_h.

The engine speed at maximum power is divided by the gear ratio to give the rotational speed of the driving wheels and this is multiplied by the rolling radius of the tyres to give the forward speed at zero wheelslip, v_o. Finally, the wheelslip can be derived from the coefficient of traction (see eqn [7.8]), provided that the net thrust developed by the driving wheels and the vertical load on the axle are available.

The net thrust from the driving wheels is equivalent to the drawbar pull plus the rolling resistance of any undriven wheels. When the implement draught force is developed by a fully mounted plough, all the forces act in approximately the same horizontal plane and do not affect the weight distribution between the axles. In addition, however, it is necessary to take account of the weight transfer from the front axle to the rear axle due to the torque required to overcome the rolling resistance of the driving wheels. The dynamic weight on the rear wheels, W_r, is given by:

$$W_r = W_{rs} + Q_{RR}/l_1 \qquad \ldots [7.24]$$

Dynamic weight on rear wheels, kN = Static weight on rear wheels, kN + Torque overcoming rolling resistance of driving wheels, kN m ÷ Tractor wheelbase, m

For a rear-wheel driven tractor, the torque required to overcome the rolling resistance of the driving wheels, Q_{RR}, is:

$$Q_{RR} = W_r \times (C_{RR})_r \times r_r \qquad \ldots [7.25]$$

whereas for a four-wheel driven tractor, the torque required to overcome the rolling resistance of the driving wheels involves the front axle as well:

$$Q_{RR} = [W_r \times (C_{RR})_r + W_f \times (C_{RR})_f] \times r_r \qquad \ldots [7.26]$$

where:

$(C_{RR})_f, (C_{RR})_r$ = coefficient of rolling resistance of front and rear wheels, dim.;

Q_{RR} = torque overcoming rolling resistance of driving wheels, kN m;

r_r = rolling radius of tyre, m;

W_f = dynamic weight on front wheels, kN.

Since the dynamic weight on the front wheels is the total weight of the tractor less the weight on the rear wheels, the vertical loads on the front and rear wheels can be determined and used in eqn [7.8] to find the wheelslip. For a four-wheel drive tractor, this results in two values of wheelslip, one for each axle. An iterative procedure is then necessary to find a common value of wheelslip which is either the same, or in some fixed relationship, for the two axles. The engine-limited power is then obtained using eqn [7.12].

Slip-limited power output

The calculation of the maximum slip-limited power is rather more complex because it involves differentiating eqn [7.12] with respect to wheelslip and obtaining the maximum point where the gradient of the power output curve, dP_o/ds, is zero. For a rear-wheel drive tractor, the power output *by combining eqns [7.8]* and *[7.12] is*:

$$P_o = [W_r \times (C_T)_{max} \times (1 - e^{-k \times s}) - W_f \times (C_{RR})_f] \times v_o \times (1 - s) \qquad \ldots [7.27]$$

The maximum power becomes:

$$W_r \times (C_T)_{max} \times (1 + k - k \times s) \times e^{-k \times s}$$
$$= W_r \times (C_T)_{max} - W_f \times (C_{RR})_f \qquad \ldots [7.28]$$

For a four-wheel drive tractor, the power output is:

$$P_o = \{[W \times (C_T)_{max} \times (1 - e^{-k \times s})_{rear}$$
$$+ [W \times (C_T)^{max} \times (1 - e^{-k \times s})]_{front}\} \times v_o \times (1 - s)$$
$$\ldots [7.29]$$

and the maximum power output is:

$$[W \times (C_T)_{max} \times (1 + k - k \times s) \times e^{-k \times s}]_{rear}$$
$$+ [W \times (C_T)_{max} \times (1 + k - k \times s) \times e^{-k \times s}]_{front}$$
$$= [W \times (C_T)_{max}]_{rear} + [W \times (C_T)_{max}]_{front} \qquad \ldots [7.30]$$

where: $(C_{RR})_f$ = coefficient of rolling resistance for front wheels, dim.;

$(C_T)_{max}$ = maximum coefficient of traction, dim.;

k = rate constant, dim.;

s = wheelslip, dim.;

P_o = power output, kW;

v_o = forward speed at zero wheelslip, m/s;

W, W_f, W_r = dynamic weight on drive axle, on front axle, on rear axle, kN.

The values of the maximum coefficient of traction and the rate constant may be different for the front and rear axles of the four-wheel drive tractor. Once the wheelslip has been obtained for either the rear-wheel driven tractor or the four-wheel drive tractor (eqns [7.28] or [7.30]), the value can be used to give the power output.

Example: David Brown four-wheel drive tractor

The following input data are taken from the appropriate OECD tractor test report, except that only half of the available gears are included for convenience.

Tyre sizes = 11.2–24 (front); 16.9–34 (rear)
Static weight = 14.70 kN (front); 19.66 kN (rear)
Wheelbase = 2.14 m

Gear	L1	L2	H1	L3	H2	H3
Engine/wheel speed ratio	134.1	80.8	67.2	46.8	40.5	23.5

Engine torque at maximum power × transmission efficiency
= 198.5 N m
Engine speed at maximum power = 2200 rev/min

Assuming a value for the cone penetration resistance of 1 MPa and using the tyre manufacturer's data book for tyre dimensions, from eqns [7.7], [7.11] and [7.24]:

Dynamic weight   = 13.91 kN (front); 20.45 kN (rear)
Mobility number, dim.   = 14.98 (front); 18.36 (rear)
Rolling resistance   = 0.95 kN (front); 1.32 kN (rear)

The engine-limited power is determined from eqns [7.8], [7.9] and [7.10]:

Maximum coefficient of traction, dim. = 0.735 (front);
                                        0.780 (rear)

Maximum total thrust   = 26.17 kN
$[k \times (C_T)_{max}]$ = 5.75 (front); 5.96 (rear)
Rate constant, dim.   = 7.82 (front); 7.64 (rear)
Thrust of front wheels, kN = $10.22 \times (1 - e^{-7.82 \times s})$
Thrust of rear wheels, kN   = $15.95 \times (1 - e^{-7.64 \times s})$

The slip-limited power output is given in Table 7.4 as the first step in the calculation of the highest available drawbar power in any gear.

The slip-limited power output is dependent on eqn [7.30]:

Wheelslip for maximum power output = 24.85%
Thrust = 8.76 kN (front); 13.56 kN (rear)
Total thrust = 22.32 kN
Total torque at driving wheels = 18.17 kN/m

Using these data, the engine-limited power output is given in Table 7.5 as the second step in the calculation of the highest available power in any gear.

By combining the relevant data from Tables 7.4 and 7.5, the highest drawbar power available in each gear, the forward speeds and the equivalent rates of work at an arbitrary ploughing performance of 0.025 ha/kW h are readily identified (Table 7.6). It is evident from this example that a similar drawbar power is available from two gear ratios and, where operationally possible, the higher speed is preferable because the wheelslip is less.

7.6 LEAST COST TRACTOR/MACHINE COMBINATIONS

With a detailed understanding of both the drawbar power

Table 7.4 Engine-limited power output in various gears as the first step in the calculation of the highest available drawbar power for the tractor specified in section 7.5.3

Gear	L1	L2	H1	L3	H2	H3
Driving torque at max. engine power, kN m	26.62	16.04	13.34	9.29	8.04	4.66
No-slip forward speed, km/h	4.57	7.59	9.12	13.10	15.13	26.08
Total thrust, kN	33.75	19.43	15.78	10.30	8.61	4.04
Wheelslip, %	>100	17.60	11.99	6.49	5.20	2.18
Travel speed, km/h	0	6.25	8.03	12.25	14.34	25.51
Drawbar power, kW	0	33.73	35.02	35.05	34.30	28.63

(*Source*: Dwyer, 1984)

Table 7.5 Slip-limited power output in various gears as the second step in the calculation of the highest available drawbar power for the tractor specified in section 7.5.3

Gear	L1	L2	H1	L3	H2	H3
Engine torque, Nm	135.51	224.90	270.42	388.29	448.69	773.28
Engine speed, rev/min	2260	1440	Above maximum torque			
No-slip forward speed, km/h	4.69	4.97				
Drawbar power, kW	21.85	23.16				

(*Source*: Dwyer, 1984)

Table 7.6 The highest available drawbar power output in various gears for the tractor specified in section 7.5.3, by combining Tables 7.4 and 7.5

Gear	L1	L2	H1	L3	H2	H3
Drawbar power, kW	21.85	33.73	35.20	35.05	34.30	28.63
Forward speed, km/h	3.52	6.25	8.03	12.25	14.34	25.51
Equivalent rate of work, ha/h	0.546	0.843	0.880	0.876	0.855	0.716
Equivalent implement width, m	1.55	1.35	1.10	0.72	0.60	0.28

(*Source*: Dwyer, 1984)

demand of the implement and the drawbar power available from the tractor, it is relatively simple to select correctly matched tractor/machine combinations. There are, however, a large number of matched combinations for any particular soil condition. Different power sizes of tractors may pull machines of various widths by adjusting the operating speed. The final choice involves not only a technical analysis but also an economic appraisal. Whilst the technical requirements can be considered for a single operation, the complete picture for tractor operating costs depends on the annual usage rather than an hourly charge. In addition. maximum profitability of the whole farm is not necessarily synonymous with minimum machinery costs for a single activity. Despite these limitations, it is worthwhile to consider the factors affecting machine width and speed for a single operation before attempting to optimise the mechanisation system for a more complex series of operations, involving a time schedule with delay penalties in the form of crop losses.

7.6.1 Least cost ploughing

The rate of work for a single operation can be increased either with *wider machines*, or with *faster speeds*, or with a combination of the two.

Wider machines

Scaling up the size of a tractor/machine combination presents few engineering problems because no new technology is required. With a minor increase in field size, field efficiency should not be adversely affected (see section 3.3). Remaining at the same travel speeds creates no new functional problems in the field, but larger machines require greater frame flexibility to maintain contour-following characteristics and are more difficult to transport along public highways. Although additional costs are incurred in bigger machine frames, stronger hitches, and heavier tractors, purchase prices remain directly proportional to width (see section 4.2).

Faster speeds

In order to contain labour costs, there is a continuing trend towards higher operating speeds, despite the need for both harder materials to combat the accelerated wear of soil-engaging elements and impact protection devices. Operator ride and implement control also become more critical items at higher speeds. Even so, the utilisation of increased tractor power through higher travel speeds results in smaller, more manoeuvrable tractors which are more adaptable to other field operations with a much lower power requirement.

Optimisation

The lowest cost is really the point of balance between the higher investment and labour costs for slow speed operations, and the higher variable costs at increased speeds. Various computer programs have been developed to identify the optimum combination of machine width and speed for a given set of technical and economic data. For rear-wheel driven tractors with an annual use of 400 h, in addition to that required for ploughing 160 ha to a depth of 20 cm, the least cost system is a 94 kW tractor operating at a travel speed of 8.3 km/h with a semi-mounted, eight furrow plough nearly two metres wide (Fig. 7.21). The equipment is replaced after 10 years, the interest rate is 8 per cent, the fuel cost is £0.024/litre and the labour charge £0.90/h. An average soil strength with a cone penetration resistance of 700 kPa is also assumed.

Perhaps rather more important than the minimum cost point are the cost contours surrounding the least cost and the sensitivity of the least cost point to price changes for equipment, fuel and labour. The cost contour in Fig. 7.21 represents the total operating costs of 5 per cent above the minimum. The large number of combinations within that boundary demonstrates the relative insensitivity of the total cost to changing machine widths, travel speeds or tractor power. A high powered tractor at 12 km/h can have a similar total cost to a much smaller tractor at 5 km/h under the particular set of circumstances. Increased labour charges shift the least cost point towards more powerful tractors, whilst higher fuel prices encourage the adoption of slower operating speeds.

Although the general conclusions from the foregoing study remain valid, the cost data for equipment, labour and fuel are very low by current standards. In a more recent analysis of least

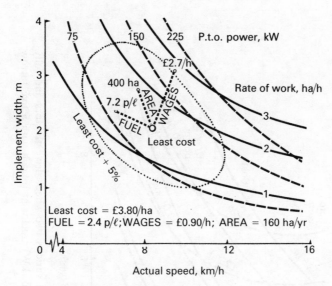

Fig. 7.21 The effect of machine size and speed on ploughing costs (after Zoz, 1974).

cost ploughing with different combinations of two-wheel drive tractors and fully mounted ploughs, the power size is restricted to a maximum of 90 kW because a second driven axle is required thereafter for efficient operation of larger tractors. The depth of ploughing is increased to 25 cm, so that fewer plough bodies are required to match the tractor power available. The present annual costs for the equipment are based on a resale age of six years and the inflation, investment and loan rates are 5, 8 and 11 per cent, respectively (see section 4.6). As the tractor power size is increased for a fixed area, the annual use is proportionately reduced from the ploughing time plus 750 hours for a 45 kW tractor to the ploughing time plus 450 hours for a 90 kW tractor. This has the effect of increasing the hourly fixed costs for the larger tractors. The repair costs for the ploughs are also speed related. This overcomes the anomaly that the faster the ploughing operation, the fewer the hours required and the lower the repair costs. For a fuel cost of £0.18/ℓ and a labour charge of £5/h, the cost contours in Fig. 7.22 indicate that high speed ploughing saves time and money. The top and side of the graph

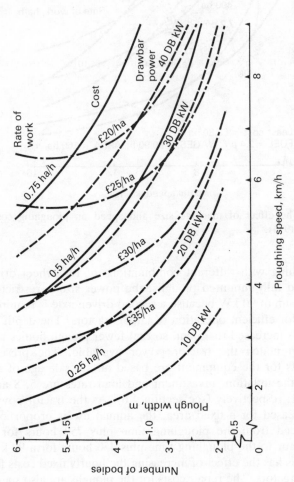

Fig. 7.22 The cost of ploughing an area of 100 ha with various two-wheel drive tractors and fully mounted ploughs is least for a 40 DB kW (about 75 p.t.o. kW) tractor at 6 km/h with a five furrow plough.

both represent practical limits for the size of fully mounted ploughs and the size of two-wheel drive tractors, repectively. The best combination is a 75 p.t.o. kW tractor and four furrow plough, operating at 6 km/h and developing just over 40 kW at the drawbar with an operating cost of £20/ha, including the hourly labour charge. Smaller tractors are more expensive than this because they cannot achieve the output, whilst bigger two-wheel drive tractors are more expensive because of the poorer tractive efficiency. A complete cost contour requires an extension of the analysis to consider four-wheel drive tractors and semi-mounted ploughs, the higher capital costs only being justified by a much larger area of ploughing.

7.6.3   Least cost cereal establishment

Instead of considering the cost of a single tractor and plough in isolation, a fleet of equally powered tractors an be selected to plough, cultivate and sow an area of winter cereals. The period for these operations cannot begin until the previous crop has been harvested and, if yield losses are to be avoided, should finish at about the optimum date for sowing the next crop. Within this timespan, the number of workdays is influenced by the minimum soil strength required – the drier the soil, the less time available. For simplicity, the three operations are completed sequentially and only one tractor is used for cultivating and sowing, regardless of the tractor fleet size.

The total cost curves for cereal crop establishment, using two-wheel drive tractors, are built up from three separate tractor/machine combinations with different machine widths in each case (Fig. 7.23). Although a fleet of three small tractors is required instead of two larger units, the main cost difference between these fleet sizes is caused by the greater penalty for untimely establishment with the smaller tractors. Additionally, of course, the higher annual wage bill for a three-man system would make its elimination more clearcut. Concentrating only on the two-tractor system, there is an optimum cost point at a tractor power level of about 80 kW. This occurs because of the increasing cost burden from the second tractor remaining idle on completion of the ploughing. Rescheduling of simultaneous operations is necessary to overcome this limitation but is unlikely to alter the least cost solution significantly.

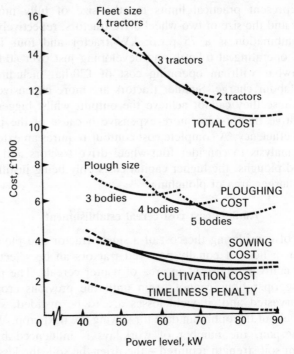

Fig. 7.23 The effect of power level of two-wheel drive tractors on the accumulated costs for three sequential operations (ploughing, cultivating and sowing) for an area of 200 ha of winter wheat (from Elbanna, 1986).

7.7 SUMMARY

Soil strength is derived from two physical properties, cohesion and frictional resistance.

The cone penetration resistance is a practical measure of the effect of soil type (clay ratio), soil moisture content and compaction level (soil specific weight) on soil strength.

Peak drawbar power is usually obtained at between 10 and 15 per cent wheelslip, depending on soil conditions.

Tractive efficiency is the ratio of tractive power output to axle power input. Like drawbar power, it is a function of wheelslip with the exception that the wheelslip for maximum efficiency is almost independent of soil conditions.

The tyre mobility number combines the main traction parameters – including soil strength, applied load and tyre configuration – into a single dimensionless ratio which determines various aspects of tyre performance such as coefficient of traction, coefficient of rolling resistance, tractive efficiency and wheelslip.

The '*Ploughing Performance Predictor*' chart provides a graphical procedure for assessing the effect of soil strength, travel speed and mouldboard shape on the drawbar power requirement for ploughing.

The dynamic weight on the rear wheels of a tractor comprises the static load, together with the weight transfer and weight addition caused by the angled draught force of the attached implement.

The '*Tractor Drawbar Performance Predictor*' chart provides a graphical procedure for evaluating the travel reduction, drawbar pull, drawbar power and ballast requirements for rear-wheel driven tractors operating on four surface conditions.

A more versatile analytical approach is available to determine the engine-limited and slip-limited power output for both rear-wheel driven and four-wheel driven tractors with fully mounted implements under any soil condition.

The selection of machinery for a single tractor operation or for a tractor fleet engaged in multiple operations involves both a technical analysis of drawbar performance and an economic appraisal of operating costs to identify a least cost solution, although the optimum system may be further influenced by whole farm considerations.

8

TRACTOR CHOICE

OUTLINE

Farmers' buying behaviour; Industrial buying; Lag and fluctuation in incomes; Personal role in purchasing; Extrovert
marketing; Dealer and brand loyalty; Adoption process and
personality profiles; Dissonance; Risk aversion; After-sales
service; Tractor power rating; Tractor performance tests;
Comparative fuel consumption; Number of gear ratios;
Lugging ability; Tractor weight to power ratio; Vehicle
flotation; Manoeuvrability; Automatic pick-up hitch; Rear
and front mounted three-point hitch; Position control;
Pressure control; Draught control with signal sensing of top
link force, lower link forces and driveline torque; Virtual
hitch point; Automatic couplers; Tractor performance
monitors; Electronic controls; Tractor automation; Parts
prices and fitment times; Forecasting tractor design feature
packages.

APOSTROPHE — HOMAGE TO THE CORN-GODS

Land belonging to the community was once known as Lammas
Land. Often, this land was let to individual farmers and fenced
off each year for growing winter wheat or Lammas wheat which
was the first corn to ripen.

The corn had to be cut on July 31st before the land reverted
to the community on August 1st, Lammas day.

Long before the advent of Christianity, the loaves baked
from the first cut corn were offered on the altar to the Celtic
Corn-god, Lugh Longhand, as a thanksgiving for the first fruits
of harvest. The word 'Lammas' is believed to have been
derived from the Anglo-Saxon *halfmaesse* meaning 'loaf mass'.

The ancient ceremonies of harvest home are just as much a part of the worship of the Earth Mother who gave life to the seed committed to her. Reverence was shown to the last part of the uncut corn by plaiting a few of the growing stalks together, for the Last Sheaf embodied the Corn Spirit and no man wished to endure the ill-luck of killing her. The hand-reapers stood round the Last Sheaf in a wide half circle and threw their sickles at it, so that these stalks were cut in common and with shared responsibility.

From the corn of the Last Sheaf was made the figure of the Corn Dolly, known in Scotland as the Maiden if the harvest was early or as the Cailleach (Old Woman) if the harvest was late. The intricately woven design with hanging ears was carried home in triumph. After occupying the place of honour at the Harvest Home feast, the Corn Dolly was kept by the hearth until ploughing was resumed after the winter solstice. She was then taken back to the fields and ceremoniously buried beneath the first furrow as an exhortation to the continued fertility in the forthcoming season.

(Sources: Humphreys 1984; Hole 1978)

8.1 INTRODUCTION

There is a wide choice of tractors and machines on the market. In some cases, almost identical equipment is dressed in different liveries so that a number of suppliers can extend their product ranges and thereby retain consumer loyalty to a particular marque. Whilst the colour of the paint may infer other brand attributes, such as reliability, the only real discriminator is price. More frequently, however, the supplier manufactures most of his own range of products, each incorporating some special features to bestow a degree of individuality to the equipment and to give marketing opportunities. Identifying the 'best buy' involves the consumer not only in bargaining, but also in evaluating the technical specification and in rating the various economic, technical, social and psychological factors in order of importance. This rating may vary from one class of machine to another, from person to person, from area to area and from country to country.

When selecting a medium-powered tractor from any of the nine leading manufacturers, one farmer survey indicated that the *main reasons for purchase*, in order of priority, were:

1. **performance;**
2. **dealer reputation;**
3. **value for money;**
4. { **dealer proximity,**
   specification,
   previous ownership of same make.

The list of reasons is of greater significance than the order of priority, for the difference in level of importance between individual items is often small, as indicated by the equal weighting given to the last three factors and their interaction with more highly rated factors.

Performance is affected by the specification of, for example, normally aspirated or turbocharged engines, manual or power shift transmissions, articulated or co-ordinated steering, pick-up hitches and automatic implement hitching. Hardware selection is impossible without a basic awareness of what these different features can offer.

The distinction between dealer reputation and dealer proximity is becoming more pronounced because the rapid reduction in annual servicing times for vehicles has allowed consumers to forsake the local dealer and take advantage of the financial savings available through major dealerships or discount chain stores.

The discount on the purchase price, the trade-in allowance and the availability of cheap finance have a greater influence on buying behaviour than variations in the annual operating costs. There are, however, significant differences in fuel economy and in spares prices which appear to have surprisingly little impact on brand loyalty.

8.2 FARMERS' BUYING BEHAVIOUR

Farmers' purchase decisions are a part of *industrial buying*, rather than of *final consumer purchasing*, because they are concerned with the accumulation of capital equipment. Industrial buying behaviour is more strongly influenced by technical characteristics, after-sales service, negotiation and bargaining, whereas final consumer buying behaviour places a greater emphasis on the social and psychological elements of the marketing mix. Although economically rational industrial buying decisions and irrational psychological consumer choices may represent the polar extremes, all purchases involve a blend of influences which are conditioned to varying degrees by personal attitudes and images.

No businessman wishes to draw attention to his susceptibility to influences which might not be considered entirely rational. Equally, farmers project an image of independence, rationality and economic motivation to the extent that the perception of a 'good farmer' is one of farming ability and economic success, rather than of community values. Farmers' desire for social status through economic success is reflected in their purchases of over-sized farm durables, such as tractors and self-propelled combine harvesters. This prestige motive may be so strong that it encourages farmers to replace equipment earlier than would be indicated by technico-economic considerations alone. This, in turn, leads to investment in machinery which exceeds the economic optimum.

Marketing organisations recognise a number of special influences which distinguish farmers, as a group of industrial purchasers of durable equipment, from other consumer groups. These special influences are:
○ **lag and fluctuation in incomes;**
○ **personal role in purchasing;**
○ **resistance to extrovert marketing;**
○ **dealer and brand loyalty;**
○ **adoption behaviour;**
○ **information sources;**
○ **investment appraisal and risk aversion;**
○ **after-sales service.**
These aspects of purchasing behaviour are studied in turn.

8.2.1 Lag and fluctuation in incomes

The lag between movements in farm income and the purchase prices of farm machinery may be up to 16 months, since farm prices are fixed annually and are less sensitive to changes in the inflation rate compared with industrial costs. Whilst this lag in farm incomes makes it easier to forecast machinery demand, income fluctuations through vagaries in the weather create unpredictable distortions in product sales. Occasional windfall profits result in a surge in machinery demand which is beyond the production capacity of the home-based industry and is satisfied at the expense of greater import penetration; whereas, any downturn in farm profitability is reflected in a sharp fall in machinery sales and production staff redundancies.

8.2.2   Personal role in purchasing

Farmers tend to purchase their general farm supplies from the nearest source for convenience. The shopping area is enlarged for machinery because dealer reputation, brand of equipment, service policies and price exert a greater influence on buying behaviour than mere convenience. Since machinery distributors provide a major source of credit to the farmer, it is not surprising that the personal reputation of the dealer for honesty and friendliness has a strong influence on dealer selection. The brand of equipment is most important to farmers who already own a large amount of equipment, the price variable is most important to middle-aged farmers, and the service variable is an underlying trend for all farmers.

The purchase of a piece of farm equipment is a very personal decision because the farmer chooses the machine to suit his own particular needs. This is usually achieved by making minor adjustments to the broad specification which has been proved to be satisfactory for the local conditions. Farmers know what they want to buy and want to know the dealer involved.

8.2.3   Resistance to extrovert marketing

Farmers are highly suspicious of advertisers' claims and of 'extrovert marketing' which is believed to hide inherent deficiencies in the product. They also totally disregard advertising which quotes 'output' figures and blame this attitude on the manufacturers who, they claim, have abused this more objective form of appeal. As a result, close scrutiny of sales literature reveals that manufacturers concentrate on selling the virtues of the product in terms of reliability and the range of options which are available to meet all needs.

8.2.4 Dealer and brand loyalty

The level of dealer and brand loyalty is quite high for farm supplies and there is a reluctance towards switching either dealership or brand too often. Dealer loyalty tends to decline with experience and the high degree of brand loyalty at any given time also decreases over the accumulation period of the machinery inventory which ultimately may comprise several different brands.

Only a small socio-economic group of farmers can be separately identified with low brand loyalty. Increased shopping activities tend to be associated with higher gross incomes, higher levels of education, larger farm sizes, and more expensive purchases. Apart from this small group, most farmers prefer not to spend time actively comparing alternatives before making a purchasing decision but the extent of shopping for machinery is greater than for other farm supplies. Although there is no relation between length of time shopping and either brand loyalty or size of investment, the average time that a farmer may be in the market for farm machinery is about a month, during which time he may have two or three contacts with each of at least two dealers. This apparent lack of interest in 'shopping around' can be partially explained by the fact that farmers generally do not perceive much difference between alternative suppliers – an inference disputed by the dealers!

8.2.5 Adoption behaviour

The *adoption process* by which people accept new products and practices involve a series of four stages:
○ **awareness;**
○ **interest– information search,**
                feasibility study,
                conviction,
                decision;
○ **action – trial,**
                adoption;
○ **confirmation.**

At the *awareness* stage, some new idea or product is noticed but there is a lack of detailed information about it. During the *interest* stage, the necessary *information* is obtained and followed by a *feasibility study* which, if favourable for the particular circumstances, leads to the *conviction* that the idea is sound and the *decision* for its adoption. The *action* stage involves a *trial* in which the new product or idea is tried on a small scale, before complete *adoption* of the innovation as a part of current practice. Thereafter, *confirmation* is sought that the decision for adoption was correct and that the results are satisfactory.

The period taken to complete the adoption process is very varied and provides a framework for distinguishing between six different *personality profiles* in any occupation, namely:

o **innovators;**
o **early adopters;**
o **early majority;**
o **late majority;**
o **laggards;**
o **non-adopters.**

The *innovators*, making up something over 2 per cent of the population, adopt new ideas when they are first introduced (Fig. 8.1). These are followed by the *early adopters*, a highly respected group of opinion leaders who are still in the van but wait until the innovation has proved its worth. Thereafter, a sizeable group which adopts the idea at a somewhat later stage can be divided into the early majority and the late majority. The *early majority* are more cautious and tend to follow the lead of the early adopters after a longer deliberation period, whilst the *late majority* regard new ideas with scepticism and only accept the change out of economic necessity or through fear of

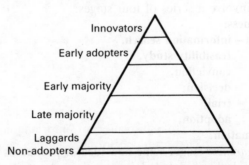

Fig. 8.1 Personality profiles of groups aware of innovation.

becoming isolated. The point of reference of the *laggards* is the past. They are the last to adopt a practice, by which time it may already be out of date, whilst a small group of *non-adopters* never accept the idea at all.

The length of the adoption period also varies substantially with the innovation. The process is usually triggered by the recognition of a machinery replacement need, a change in farm operations, desirable changes in new models, or savings in labour requirements. Simple ideas, easily tested, compatible with previous experience and yielding measurable results, are likely to be adopted more rapidly than their more complex counterparts, irrespective of the cost-benefits involved.

The innovators initiating the adoption process have more education, greater formal participation in organisations, higher social status, younger age and a greater propensity for reading. They tend to be more venturesome, more cosmopolitan and more favourably disposed towards the use of credit, perhaps linked to their ownership of large farms with high gross incomes from specialised enterprises. The adoption behaviour of the innovators indicates that they become aware of new products and practices at an earlier date relative to the average farmer by more direct contact with agricultural scientists, through research literature and from the farming press.

8.2.6 Information sources

Farmers are continuously exposed to a stream of information from numerous sources. Manufacturers and dealers actively portray the attributes of their particular products through personal contact, field demonstrations and the media. Extension personnel tend to provide farmers with more objective information, whereas friends and neighbours provide information based on actual experience with the product.

The importance of these information sources varies as the farmer proceeds through the adoption process. The trade, particularly by advertising through the mass media, is the most important and effective dispenser of information during the awareness stage. The experience of other farmers and the deliberations with members of the family are regarded as more reliable and accessible information sources for the decision-making processes during the interest stage. Advisory information from both the extension services and from dealers is most influential in resolving specific implementation problems at the action stage. Even after adoption decision, however, the information search may well continue in order to reduce *dissonance*, or inconsistency, and provide confirmation that the proper decision was made.

8.2.7 Investment appraisal and risk aversion

Farmers use few formal methods to help them determine the amount of capital which can be invested profitably in the farm business, the major precipitating circumstances being the desire to own larger equipment and the need to replace worn-out items. Only a minority base their investment on cash flow, return on investment, and a self-imposed debt limit.

The success of the investment in farm machinery is judged primarily according to its contribution to farm output and not in terms like 'satisfaction' or 'styling'. This means that farmers are wary of any major design changes and are looking for proven reliability rather than innovation.

Farmers also tend to have a negative approach to the buying decision by firstly identifying features they dislike most before those that appeal to them. This conservatism is a form of *risk aversion*. Most farmers have found a level of risk for given situations and will avoid unnecessary or undesirable risks which may jeopardise their present level of income.

8.2.8 After-sales service

The most crucial aspect of selling to farmers is their insistence on good parts and servicing operations because any delay in repairing a major breakdown during a critical field operation can incur severe penalties through loss of yield. This is likely to pose an increasing problem over the next few years as the costs of maintaining parts stores and service workshops continue to rise. Already, there is a major change in routine vehicle servicing from the traditional garage to the superstores and specialist outlets, a change made possible by the extension in service intervals. A car which required a service time of 24 hours every 15 000 km only a few years ago now needs only four hours.

8.3 POWER RATING

The key feature in the choice of a tractor from the huge range available is the power rating. As the demand for tractors is very much within a replacement market, the buyer knows the power requirements for his existing machinery and any limitations that may have occurred. These limitations may justify changing to a more powerful tractor, the extent of the power increase being more dependent on salesmanship than on precise need. In addition, high labour costs continue to encourage the use of fewer tractors with a higher power rating. This means that the replacement tractor is initially always over-powered and under-utilised for the work in hand, until the size of the replacement machinery redresses the balance.

Direct replacement of one tractor with another of the identical power rating presents more of a marketing problem because there is the danger that the performance of the new tractor, whilst it is being run-in, is inferior to the one that it is replacing. In order to minimise adverse consumer reaction, part of the marketing strategy involved in frequent model changes is upgrading the power rating by some 5 per cent. Although the increase is small, it does gradually mount up and it is very rare for the customer to revert to smaller machines!

8.3.1 Performance standards

Whilst the importance of tractor power rating is not in dispute, the way in which it is measured and presented can lead to considerable confusion. The anomalies occur through the use of differing standards for measuring engine power output. The three standards commonly used for quoting diesel engine power in agricultural machines are the British Standard BS AU 141a: 1971, the German (Deutsche Industrie Norm) DIN 70020 and the American (Society of Automotive Engineers) SAE J816a.

Power at the engine flywheel, although traditionally used as the measure of tractor power in Europe, is really only of academic interest. The engine *brake power*, so named because it was measured by means of a 'brake dynamometer', is the power available from an engine, sometimes tested without essential accessories (a bare engine test) and without any smoke restriction.

Before the tractor can be put to work, some of the engine power developed in a bare engine test must be used to drive the alternator and cooling fan, the engine may have to be de-rated for the vehicle to comply with the legislated limit on the emission of exhaust smoke, and the engine power must be transferred to a power outlet point involving losses through a mechanical, hydraulic or electrical transmission system. The mechanical transmission system is by far the most efficient and the proportion of the engine power lost in the mechanical drive system to the tractor power take-off can be expected to be similar on different types of tractor of comparable size. For this reason, the *power take-off power* is a fair indication of the maximum power available from the tractor for other functions as well as for driving attached machines. It should be noted that when a tractor is fitted with a multi-speed power take-off, the maximum power is often slightly higher at the higher power take-off speed because of the slightly better gear efficiency.

For uniformity of testing conditions, *drawbar power* is only measured on concrete. Whilst drawbar test results for different

Table 8.1 The wide range of power ratings for tractor engines which give similar levels of performance at the power take-off and at the drawbar

Tractor make and model	Claimed engine power, kW	Official max. power take-off power, kW	Official max. drawbar power, kW	Testing organisation
Belarus MTZ80	67 (SAE)	56	50	Nebraska (1973)
Ursus C385A	63 (SAE)	53	49	OECD (1979)
Deutz DX85	60 (DIN)	54	47	OECD (1979)
MF590	55 (DIN)	53	47	OECD (1979)

(*Source*: Jeffrey, 1981)

tractors are useful for comparative purposes (a small difference in performance on concrete inferring an even greater difference in the field), the individual results for a particular tractor are less easily related to farm applications than the power take-off power.

In order to achieve comparable power take-off and drawbar performance in official tests, engine power outputs, as quoted by tractor manufacturers, may differ enormously depending on which standard test has been used (Table 8.1). The variation in the claimed engine power ratings, in the absence of the other data in the table, is more than sufficient to have a substantial influence on the purchasing decision.

8.3.2 Tractor performance tests

The official OECD tractor test involves a series of compulsory tests of power take-off performance, drawbar performance and fuel consumption at both maximum power and under part loads. A sample of the power take-off performance and fuel consumption data from the test report for a Ford 8210 tractor is given in Table 8.2 and shown graphically in Fig. 8.2, whilst the drawbar performance data are given in Table 8.3.

The *rated engine speed* is the speed at which full engine power can be developed continuously. When the rated engine speed is greater than those corresponding to the standard power take-off speeds of either 540 rev/min or 1000 rev/min, there will be less than the maximum power available at the power take-off. This is important because equipment manufacturers design their machines to conform to the standard power take-off speeds.

The maximum drawbar performance is not obtained in the lowest gear because the high axle torque available leads to wheelspin. As more of the engine power is utilised at higher travel speeds, so drawbar power reaches a maximum. Even on concrete, this peak drawbar power is some 10 to 15 per cent below the maximum power take-off power available.

Fuel consumption increases as more power is developed but the overall efficiency of the engine also improves and reaches a peak at around 90 per cent of maximum power at rated engine speed. Low fuel use in itself is of no relevance because it can be achieved only at the expense of a low power output and a slow

Table 8.2 Test data on the power take-off performance of a Ford 8210 tractor

| Power, kW | Speed, rev/min | | Fuel consumption | | |
| | Engine | p.t.o. | Hourly | | Specific, kg/kW h |
			ℓ/h	kg/h	
Maximum power – 2 hour test					
74.7	2300	1122	24.90	21.14	0.283
Maximum power at the rated speed					
74.7	2300	1122	24.90	21.14	0.283
Maximum p.t.o. power at the standard p.t.o. speed of 1 000 rev/min					
71.0	2049	1000	22.52	19.09	0.269
Parts loads					
(1) Power for 85% of torque obtained at rated speed					
64.6	2338	1141	22.74	19.28	0.298
(2) Power for three quarters of torque defined in (1)					
48.9	2358	1151	18.51	15.70	0.321
(3) Power for half of torque defined in (1)					
32.8	2373	1158	15.03	12.75	0.388
(4) Power for one quarter of torque defined in (1)					
16.4	2383	1163	11.68	9.91	0.604
(5) Unloaded					
0	2430	1186	8.61	7.30	—

(*Source*: NIAE, 1985)

rate of work. Instead, the aim is low fuel use per unit of power output, that is, a low *specific fuel consumption* (kg/kW h).

8.4 FUEL ECONOMY

Fuel represents a large proportion of the total operating costs for tractors. Differences in tractor engine design lead to a significant variation in fuel consumption. This variation is more readily appreciated in terms of the annual expenditure on fuel for tractors with a similar power rating. In addition, good machine maintenance and efficient operation can save a further 10 per cent of the annual fuel costs.

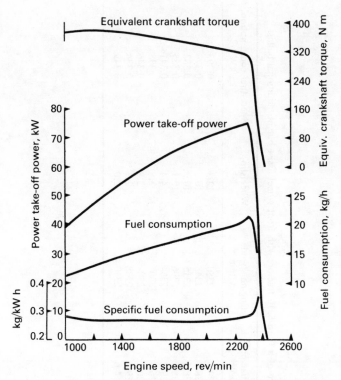

Fig. 8.2 Test data for the power take-off performance and fuel consumption of a Ford 8210 tractor (after NIAE, 1985).

8.4.1 Effect of engine design

The answer to the question of fuel economy is largely determined by the choice of engine design between the *direct injection* system which uses a spray nozzle injecting fuel directly into a combustion chamber formed in the piston crown, and the *indirect injection* system which employs a pintle nozzle injecting fuel into a pre-combustion chamber formed into the cylinder head. The indirect injection system has better cold starting characteristics but has been largely discarded in favour of the direct injection system which provides fuel savings of the order of 13–14 per cent.

Table 8.3 Test data on the drawbar performance of a Ford 8210 tractor, with front wheel assist but no additional ballast

Gear*	Speed, km/h	Drawbar power, kW	Drawbar pull, kW	Engine speed, rev/min	Wheelslip, %	Spec. fuel consumption, kg/kW h
2L	4.10	44.5	39.0	2350	12.3	0.393
3 LM	4.34	46.8	38.9	2348	10.6	0.376
3L	5.30	56.1	38.3	2315	10.3	0.382
4LM	5.87	62.1	38.1	2306	10.1	0.355
4L	7.95	64.4	29.2	2304	4.4	0.341
1HM	8.88	65.5	26.6	2307	3.8	0.338
1H	11.58	64.7	20.1	2304	2.8	0.346
2HM	12.29	67.3	19.7	2308	2.7	0.334
2H	15.96	64.5	14.6	2310	2.0	0.349

*L = low range; H = high range; M = torque amplifier engaged
(*Source:* NIAE, 1985)

The type of engine cooling system also affects the operating temperature of the engine and, hence, its thermal efficiency in converting fuel into useful power output. Most engines are *water cooled* but *air cooled* engines can provide a fuel saving of 5–7 per cent at maximum power. At part loads, the advantage declines because the difference in the running temperature for the two cooling systems is less.

The use of *turbochargers*, as a means of increasing the engine power output, has more mixed effects on fuel economy. A turbocharger is an exhaust gas driven turbine which is directly coupled to a centrifugal compressor forcing a greater charge of air into the engine cylinder. As it is the volume of air that limits the peak performance of an engine (extra fuel being easily injected at a cost), an additional charge of air increases the maximum power by as much as 30 per cent. It also effectively increases the combustion pressure of the gases within the engine cylinder and can give more efficient use of fuel but only for a particular part load when the air/fuel ratio is near to the optimum. At low engine speeds, turbocharging hardly alters the specific fuel consumption compared with normal aspiration; whilst at peak power, turbocharging can give 10 per cent better fuel economy but only by avoiding fuel wastage. The black exhaust smoke which is often taken as representing an engine 'working hard to earn its keep' is simply an indication of the amount of unburnt fuel being thrown away.

8.4.2 Comparative fuel demand

At first glance, the differences in the specific fuel consumption figures appear unimportant (Table 8.4). Indeed, tractor fuel use is often compared on the basis of the number of tank refills required, with scant regard to the different tank capacities! The final reckoning, however, is the annual fuel account.

The annual fuel expenditure, for diesel at 20 pence/litre, varies by about 20 per cent for the different tractors (Table 8.4). The costs are based on an annual use of 1000 hours at a part engine loading of 55 per cent of the maximum power take-off power, for which the specific fuel consumption is taken as similar to that at maximum power – in the absence of more detailed figures.

Table 8.4 Annual fuel costs for medium power tractors

Tractor make and model	Max. p.t.o. power, kW	Spec. fuel cons. at max. p.t.o. power, kg/kW h	Fuel cons., ℓ/h	Annual fuel cost, £/yr Total	Annual fuel cost, £/yr Per kW of max. power
Deutz Fahr DX 3.90	54.5	0.227	14.9	1640	24.97
Fiat 80–90	56.6	0.243	16.6	1823	26.73
Renault 781S	52.8	0.266	16.9	1861	29.26
Massey Ferguson 590	53.6	0.275	17.8	1953	30.25

Although the annual expenditure comparison is not strictly accurate because of the different power ratings of the individual tractors, the rather less practical but more precise annual fuel cost per kilowatt of maximum power yields the same conclusion.

8.5 POWER TRANSMISSION

Agricultural tractor operations may be classified into four *transmission speed ranges*:

- **creep speeds;**
- **slow field speeds;**
- **fast field speeds;**
- **road speeds.**

In the *creep speed* range between 1 and 3 km/h, one or two speed ratios are required for specialised transplanting and difficult harvesting operations. Heavy draught and power take-off operations, such as ploughing, cultivations and forage harvesting, require a minimum of three, and preferably four, speed ratios to adequately cover the *slow field speed* range between 3 and 10 km/h. For lighter work, including mowing, swathing and spraying, two speed ratios are sufficient to span the *fast field speed* range from 10 to 20 km/h. In the highest range for *road speeds* up to 32 km/h, it is preferable to have two or three speed ratios to match varying transport conditions – fully-loaded or empty trailers, farm tracks or metalled roads – but the average loading is generally below that for the slow field speed range and the hours of use are less. On this basis, 8 to 10 gears are justified.

As the spread widens between the minimum and maximum speeds, a greater the ratio coverage is needed in the gearbox. A tractor/implement combination may be expected to work effectively at plus or minus 20 per cent of the ideal speed. It follows, therefore, that the constant *geometric spacing factor* should not exceed 1.4 for a set of successive speed ratios in a mechanical gearbox. The number of speed ratios to cover the total speed range is given by the expression:

$$V_{max}/V_{min} = k_g^{(N-1)} \qquad \ldots [8.1]$$

Maximum speed, km/h/Minimum speed, km/h = (Geometric spacing factor, dim.)$^{(\text{Exponent related to Number of ratios})}$

Applying this constant geometric spacing factor of 1.4, a tractor with a maximum speed of 32 km/h in the highest gear, a minimum speed of 1.5 km/h in the lowest gear (both at the rated engine speed) should be fitted with a 10-speed transmission system to provide overlap between adjacent gear ratio.

Economy in gearbox design demands that such a plurality of speeds can only be provided by using a 'progressive range system' wherein the ratios in each range sequentially follow the ratios in its adjacent range. For example, a manually operated high/low ratio change in conjunction with a basic three speed conventional gearbox provides six forward and two reverse speeds. This can be further compounded by the addition of an hydraulically actuated 'on-the-move' ratio change, thus giving twelve forward and four reverse speeds. 'On-the-move' gear changes which avoid the need to stop the tractor before attempting to change to a higher gear save time and increase operator productivity. From the buyer's point of view, the inherent danger of compounding the gear ratios is that some gears cover almost the same speed ranges, leaving undesirable gaps between the speed ranges of adjacent gears elsewhere.

A typical spread of gears is shown in Fig. 8.3. In the lower gears below 5 km/h, wheelslip sets the limit to the pull which can be delivered by the tractor. As full engine power is not developed, the fastest wheel speeds in these gears are governed by the rated engine speed and slower wheel speeds are obtained by throttling back the engine speed down to some 40 per cent of rated speed.

In the higher gears, however, the pull may be sufficient to bring the engine to full power and even to *lug* the engine speed down to its stalling speed (see Fig. 8.2).

- **Lugging** is the ability of an engine, already delivering full power at its rated speed, to meet an extra load which causes the engine speed (and so the wheel speed) to fall, whilst the torque output of the engine rises.

Lugging can continue down to the engine speed where the engine gives its maximum torque (usually at about two thirds of the rated speed), but below that speed, the engine will stall. In practice, few operators would allow the engine speed to fall that far, and few manufacturers would recommend lugging for any length of time. After the engine speed has been lugged down towards

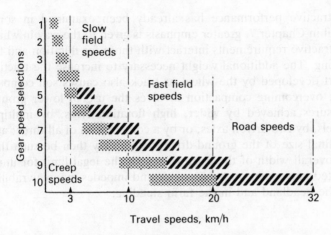

Fig. 8.3 The travel speeds from 40 per cent of full engine speed up to rated engine speed for a range of overlapping gear ratios, with the lugging range in the higher gears shown hatched.

the maximum torque, the only way of meeting a further increase in load is by changing down to the next lower gear without too great a drop in speed.

As well as the correct overlapping of the gear ratios, proper location of the transmission controls and the ease of the gear change also contribute to operator productivity. When the controls are in a difficult position or when the physical effort to make the change is excessive, speed changes are avoided even though the tractor has the potential for more efficient operation by frequent transmission speed changes. This aspect of tractor choice is best investigated in a field demonstration of performance in combination with the machinery which impose particular operational requirements on the tractor.

8.6 GROUND-DRIVE SYSTEMS

The ground-drive system for a tractor must fulfil three requirements, namely:

○ **traction;**
○ **flotation;**
○ **manoeuvrability.**

As tractive performance has already been examined in some detail in Chapter 7, greater emphasis is given to the way in which the tractive requirements interact with those for flotation and for turning. The additional weight necessary to increase the tractive effort developed by the wheel or track also causes soil compaction; overcoming compaction involves the use of lower ground pressures achieved by wider, high flotation tyres, by multiple wheels, by two driven axles, or by a combination of all three; and the final size of the ground-drive system may then be such that the overall width of the tractor exceeds the legal limit for unrestricted movement on public roads and impedes manoeuvrability on the headland and in the farm steading.

8.6.1   Two-wheel drive and four-wheel drive tractors

Drawbar pull alone is not the most important factor in the choice of tractor ground-drive system because maximum drawbar pull is only developed at a high wheelslip which reduces the travel speed. For the maximum drawbar power output in the field, it is important to achieve the correct balance between the engine power available, the travel speed and the weight on the driving wheels. The optimum weight on the driving wheels per unit of available power varies with the travel speed – the lower the speed, the greater the weight required (Fig. 8.4). For this reason, a turbocharged tractor is more suitable for power take-off duties because it is only likely to have sufficient weight to fully utilise the extra power at the drawbar in good conditions, as indicated by the lower curve in Fig. 8.4.

The line marked '4WD' shows that four-wheel drive tractors, when fitted with tyres which are capable of carrying a mass of 100 kg/kW (the normal fitment), can achieve maximum tractive efficiency at a speed of 6.5 km/h. Two-wheel drive tractors are usually fitted with tyres which are capable only of carrying up to about 65 kg/kW and need to be operated at a speed of nearer 10 km/h for optimum tractive efficiency, although performance may be acceptable in good tractive conditions down to a speed of 7 km/h.

The greater weight of four-wheel drive tractors is largely responsible for their ability to achieve higher rates of work on

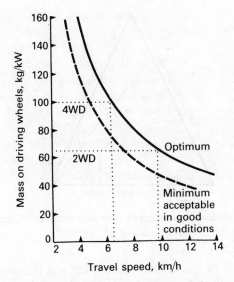

Fig. 8.4 Mass on the driving wheels required for maximum tractive efficiency at different speeds (after Dwyer, 1982).

heavy draught operations than two-wheel drive tractors of the same power. The two-wheel drive tractor is unlikely to reach its optimum speed of nearly 10 km/h, particularly in poor tractive conditions. Consequently, differences in work rates of up to 30 per cent in favour of four-wheel drive tractors with equal sized wheels all round, and 15 per cent in favour of front-wheel assist tractors (equipped with smaller drive wheels on the front axle compared with those on the rear axle) can be generated.

Although there is some scope for improving the work rates of two-wheel drive tractors by fitting larger sizes of tyres, their availability and cost limit the engine power to about 60 kW for maximum efficiency. Front-wheel assist tractors can operate effectively up to about 100 kW, and four-wheel drive is necessary thereafter.

8.6.2 Vehicle flotation

Soils are not usually strong enough to resist traffic loads without some tracking and sinkage. A rut is formed when the ground

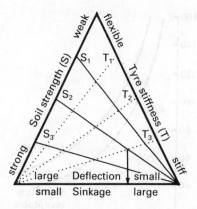

Fig. 8.5 The relationships between soil strength, tyre stiffness, tyre deflection and sinkage (after Soane *et al.*, 1980/81).

contact pressure of the vehicle exceeds the soil bearing strength. The ground contact pressure of a pneumatic tyre is similar to the tyre inflation pressure, although localised higher ground contact pressures also occur due to the effect of additional tyre stiffness in the sidewalls and lugs. The deflection and sinkage of tyres are related to both tyre stiffness and soil strength but in opposite ways (Fig. 8.5). The qualitative extent of tyre sinkage is defined along the base of the diagram at a point vertically below the intersection of the two lines representing any particular combination of tyre stiffness and soil strength. Thus, sinkage becomes greater as tyre stiffness is increased through the higher inflation pressures required to carry bigger wheel loads as the alternative to changing to larger tyres. Deep ruts and excessive tracking damage the stucture of the soil to the extent that crop yields are affected.

It is not just the ground contact pressure that is important but also the total vehicle weight. As tractors have become larger, it is no longer uncommon to see dual tyres and even triple tyres fitted to spread the weight of the vehicle over a larger ground contact area. Even these multi-tyre systems are seldom adequate to achieve average ground contact pressures of *less* than those for smaller tractors; comparable pressures are not enough because these only limit the soil damage in the plough layer. The peak soil stresses under a tyre are concentrated along the centre-line

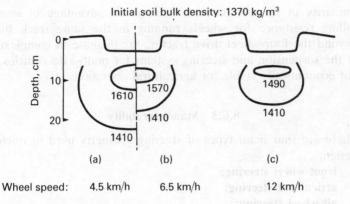

Fig. 8.6 The reduction in compacted soil density and depth of compaction with increasing wheel speed (adapted from Soltynski).

of the rut at a depth equal to half the rut width below the bottom of the rut (Fig. 8.6). Wider tyres and tracks transmit the vehicle load deeper into the soil. With tyres above 0.5 m wide, the peak soil stress may occur below the plough layer and loosening will require heavy subsoiling. Close coupled, dual tyres interact with each other and have an effect equivalent to the full width of both tyres. When dual tyres are spaced more than one wheel width apart, the point of peak stress below each rut reverts to that for the individual tyres. Although keeping the wheels apart is preferable, it does extend the width of the vehicle and imposes greater bending forces in the wheel axles.

Vehicle speed and number of wheel passes also affect the extent of soil compaction, since the dynamic response of the soil varies with the duration and number of applications of a particular stress cycle. For soil at an initial bulk density of 1380 kg/m³, the effect of the passage of a wheel at increasing speeds is to reduce both the level of the peak stress and the depth of soil compacted. Thus, high speed field operations improve vehicle flotation as well as power utilisation.

The alternative to dual wheels, side by side, is tandem wheels, one behind the other. Multiple wheel passes in the same rut cause a similar amount of soil deformation as a single wheel with a ground contact pressure equivalent to the aggregate of the contact pressures for the multiple wheels. Apart from the similarity in soil compaction, there is an advantage of lower rolling resistance for wheels running in the same track but, beyond the four-wheel drive tractor, the increase in complexity of the suspension and steering systems for multi-axle vehicles is not economically viable for agricultural operations.

8.6.3 Manoeuvrability

There are four main types of steering geometry used in tractor design:

○ **front-wheel steering;**
○ **articulated steering;**
○ **all-wheel steering;**
○ **skid steering.**

Front-wheel steering provides excellent manoeuvrability for conventional tractors with only one drive axle (Fig. 8.7(a)). Small front wheels can be turned through a large yaw angle before fouling the body of the tractor. Once extra traction is demanded

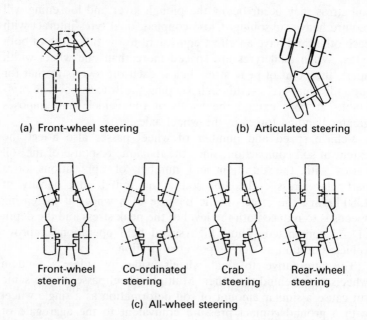

(a) Front-wheel steering   (b) Articulated steering

| Front-wheel steering | Co-ordinated steering | Crab steering | Rear-wheel steering |

(c) All-wheel steering

Fig. 8.7 Steering configurations.

from the front axle, the larger wheels which are desirable to improve grip adversely affect turning. For front-wheel assist tractors, ingenious design of the steering geometry can reduce the turning circle by over 10 per cent. In the patented 'castor action' steering system, the inclination of the wheel kingpins by 12 degrees backward has the effect of tilting the wheel on the inner radius of the turn under the chassis of the tractor when on full steering lock. The steering benefit from this design feature disappears when equal sized wheels are fitted to four-wheel drive tractors and the turning circle becomes inconveniently large. In addition, all four wheels of a front-wheel steered tractor break new ground during turns, giving comparatively poor traction in soft ground.

Articulated steering overcomes both the turning and tracking limitations of front-wheel steering by dividing the tractor chassis into two parts, each sub-frame having its own fixed axle but coupled together at a central pivot (Fig. 8.7(b)). The front half of the tractor usually incorporates the engine, gearbox and cab, whilst the rear half supports the implement hitch. The tractor is steered by large double acting hydraulic cylinders on each side of the pivot point. By locating the pivot point centrally between the two axles, the front and rear wheels track on the same turning circle.

Whilst the manoeuvrability is impressive with trailed equipment on level ground, the steering characteristics are less predictable with mounted implements. When an implement is raised out of work for a turn on the headland, the shift in weight distribution to the rear wheels makes them less responsive to articulation, so that the front frame skids sideways to make the turn. Minor steering corrections in work have the effect of swinging the attached machinery in the opposite direction to the correction. This is particularly evident on side slopes because articulating the front sub-frame uphill results in the rear sub-frame turning downhill. The horizontal position of the centre of gravity also shifts downhill and reduces the lateral stability of the tractor.

All-wheel steering of a tractor with a rigid frame offers the greatest operational flexibility (Fig. 8.7(c)). Full manoeuvrability is possible with wheels of any size and there are no steering problems on side slopes. Using all-wheel steering, there are four steering methods available:

○ **front-wheel steering;**
○ **co-ordinated steering;**
○ **rear-wheel steering;**
○ **crab steering.**

Front-wheel steering is entirely satisfactory for 75 per cent of the operating time in the field and conforms to the handling characteristics of a two-wheel drive tractor.

Co-ordinated steering is used for sharp turns, the rear wheels tracking behind the front wheels, just as for articulated steering. This means that co-ordinated steering, like articulated steering, also causes the implement to swing to the wrong side immediately after a steering correction has been initiated (compare linkage positions in the second frame of each steering pattern in Fig. 8.8). For a given travel speed, the maximum transverse deviation grows with increased rate of turn.

Rear-wheel steering is of help when hitching implements and when manoeuvring in confined spaces, whilst crab steering is used on sidlings to counteract downhill drift by angling all four wheels up the slope. With crab steering, the tractor frame remains pointing straight ahead and the lateral stability of the tractor is unaffected by the steering correction.

The only disadvantage of all-wheel steering is the additional cost for both the manufacture and maintenance of double the number of universal joints and axle pivot points compared with conventionally steered tractors.

Skid steering is employed in crawler tractors and can give a very tight turning circle. When a brake is applied to the final drive for one track, the speed of the other track increases through the action of the differential. Abuse of the turning technique can impose considerable side loads on the ground-drive system and even result in stripping a track off its drive sprocket. Skid steering is also used on lightweight wheeled vehicles to minimise design and manufacturing costs where there are more than two drive axles.

A computer simulation illustrates the differences between the steering patterns for a loop turn at the headland for three types of four-wheel drive tractors, each equipped with a rear-mounted implement such as a chisel plough (Fig. 8.8). In order to turn the outfit, the operator turns the steering wheel to the left at a constant rate, which gives a continually increasing yaw angle of

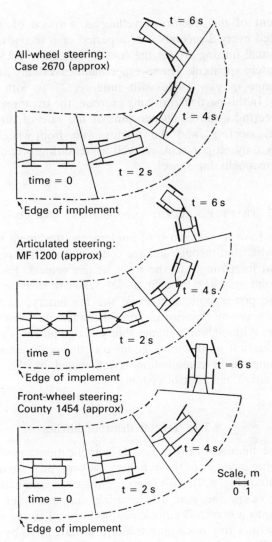

All-wheel steering:
Case 2670 (approx)

t = 6 s

t = 4 s

t = 2 s

time = 0

Edge of implement

Articulated steering:
MF 1200 (approx)

t = 6 s

t = 4 s

t = 2 s

time = 0

Edge of implement

Front-wheel steering:
County 1454 (approx)

t = 6 s

t = 4 s

t = 2 s

time = 0

Edge of implement

Scale, m

0 1

Fig. 8.8 Turning characteristics of three types of steering systems used for four-wheel drive tractors (after Mertins, 1978).

3.44 °/s at the outer wheel for both front-wheel steering and co-ordinated steering (or a continually increasing angle of articulation of 6.88 °/s at the pivot point with articulated steering). The

movement of the tractor, travelling at a speed of 9 km/h, is
monitored every 2 seconds over a period of 6 seconds.

The small turning arc for the tractor with all-wheel steering is
immediately apparent. The 'average manoeuvrability', defined as
the change in yaw angle with time, is 24 °/s with all-wheel
steering. In this particular driving exercise, the tractor with articulated steering achieves 86 per cent of the rate of turning with
all-wheel steering, and the tractor with front-wheel steering
manages only 60 per cent, as well as requiring proportionately
greater manoeuvring space.

8.7   TRACTOR/MACHINE CONNECTIONS AND CONTROL

Tractors have a wide range of implement attachment options at
the drawbar and through linkage connections to the rear, to the
side, and increasingly to the front of the vehicle. Each attachment point requires its own power take-off shaft and remote
hydraulic power outlets to drive the machinery, as well as a
complex servo-mechanism to control implement movement and
response. Although interchangeability of equipment is assured by
standardisation of the hitch geometry, the performance of the
tractor/implement combination is influenced by the choice of
hitch position and control system.

8.7.1   Tractor drawbar hitches

Even the humble drawbar has its complications, now that the
time-saving qualities of the *automatic pick-up hitch* are more fully
appreciated (Fig. 8.9). This hitch comprises an arm with one end
pivoting below the rear axle housing and the other free end
shaped into a hook. The hook can be lowered to the ground to
engage with a ring hitch on a trailer drawbar, lifted by means of
the rear linkage hydraulic system, and locked into the transport
position with a mechanical latch. As the hook cannot be seen by
the driver, the blind implement docking operation relies heavily
on driver experience and familiarity with the range of equipment.
In addition, the pick-up hitch point is much further forward than
the clevis of a conventional drawbar. This restricts the manoeuvrability of a tractor/trailer combination because the trailer

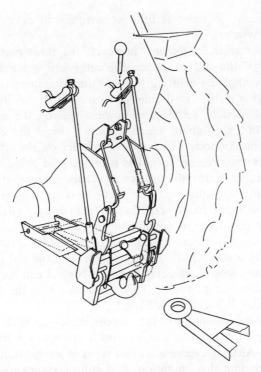

Fig. 8.9 Once the hook of the automatic pick-up hitch is positioned under the eye of the trailer drawbar, the hitch is lifted by the rocker arms of the tractor three-point linkage and mechanically latched for transport.

drawbar fouls the tractor rear tyre on the inside of the turn at a smaller angle of articulation.

Moving the hitch point towards the rear axle can also limit the tractive performance of a two-wheel drive tractor in conjunction with a rear-wheeled trailer. The tractor overturning moment caused by the vertical load at the end of the trailer drawbar effectively increases the amount of weight transferred from the front axle of the tractor onto the rear axle. As the hitch point is moved closer to the rear axle, the moment arm becomes smaller and less weight is transferred. Of course, four-wheel trailers almost completely support the weight of their own cargo and only transmit a horizontal pull through the hinged drawbar to the tractor, so there is little opportunity to develop weight transfer anyway.

For any trailed equipment, however, the three-point linkage system with the aid of *pressure control* can generate weight transfer by applying a lifting force to the equipment (or trailer) drawbar (Fig. 8.10). Hydraulic pressure within the tractor lift cylinder is used to provide an upward pull on the equipment drawbar through a chain which is looped round the equipment drawbar and connected to a special coupler on the tractor rear links. Once the pressure level has been selected by adjusting the operator controls for the tractor hydraulic system, it is maintained automatically at the set amount, regardless of tractor/equipment movement due to field irregularities. When the tractor goes over an undulation, any change in chain tension is sensed as a pressure change by the control system which responds by automatically raising or lowering the links to maintain constant weight transfer. Pressure control can increase the available drawbar power by 7 kW, provided that the front end of the tractor is correctly ballasted.

Some of the advantage of the automatic pick-up hitch is eroded by having to leave the cab to attach special weight transfer couplings. Although there are other types of weight transfer hitch which overcome this limitation, it is still necessary to dismount and manually connect brake pipes and hydraulic services for tipping trailers and the power take-off shaft for power-driven machinery.

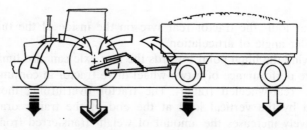

Fig. 8.10 When pressure control is used to apply a lifting force through a special coupler on the rear linkage, the chain pulls upwards on the trailer drawbar which increases the vertical load available for traction at the rear wheels of the tractor by weight addition from the trailer and by weight transfer from the front axle of the tractor.

8.7.2 Rear mounted three-point hitches

The *three-point hitch* provides an elegantly simple method of implement attachment, incorporating a hydraulic system for implement lift and for automatically controlling the implement position or the implement forces acting on the tractor. It comprises two lower links and a single top link, forming a triangular attachment for implements by means of three pivot pins. In the horizontal plane, convergence of the pivoting lower links resists displacement of the *virtual hitch point* from the centre line of the tractor; the greater the lateral disturbance, the greater the restoring moment trying to swing the implement back into line unless prevented by tightening the check chains (Fig. 8.11(a)).

- A **virtual hitch point** is the imaginary point through which the resultant hitch force is assumed to act.

The point of convergence of the top and bottom links governs the attitude of the implement during lifting and lowering movements, influences ground penetration by the implement, and also strongly affects the weight transfer and weight addition which can be developed. By careful design of the linkage geometry, the location of the vertical point of convergence alters with the angular position of the linkage, moving upwards and towards the front of the tractor as the implement is lowered into work (Fig. 8.11(b)). When there is no lifting force in the lift rods, the point of convergence of the links also represents the virtual hitch point through which the resultant force acts in the vertical plane. This is important because an initially short, low point of convergence gives a steep implement attitude and a draught force acting at a shallow angle for rapid ground penetration; whilst a long, high point of convergence allows vertical movement of the implement without a significant change in attitude and produces a steeper line of action to increase weight transfer when maximum pull is required by the fully lowered implement.

The angle of the resultant force acting on the tractor can also be varied by exerting a lifting force on the lower links. When the lower links are restrained in this way, the virtual hitch point is no longer superimposed on the point of convergence of the links and shifts towards the rear of the tractor. As the lifting force is increased, the line of action of the resultant implement force steepens and weight transfer is increased. *Draught control*

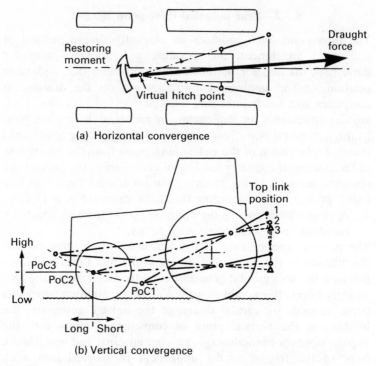

(a) Horizontal convergence

(b) Vertical convergence

Fig. 8.11 (a) Horizontal convergence of the pivoting lower links resists displacement of the virtual hitch point from the centre line of the tractor, the restoring moment to swing the implement back into line increasing with the magnitude of the lateral disturbance; (b) vertical point of convergence (PoC) of the upper and lower links alters as the implement is lowered into work, good ground penetration in position 1 being substituted for good weight transfer in position 3 (after Inns, 1985).

enables the hydraulic system to adjust this lifting force automatically and to maintain the draught force of the implement within pre-set limits by continuously monitoring the draught developed by the implement to provide the necessary control signal. This control signal can be sensed in three ways:

○ **top link sensing;**
○ **bottom link sensing;**
○ **driveline torque sensing.**

Top link sensing of the draught control signal is best suited for *fully mounted* equipment on tractors of up to about 60 kW. Instead of sensing draught force itself, the top link is more correctly sensing moments of the forces generated by implements about the ends of the lower links pivoting on the cross-shaft – the cross-shaft moments (Fig. 8.12). This works satisfactorily for small implements because the draught force generates the most of the compression force in the top link. As a tractor becomes increasingly more powerful, it is matched with a longer implement whose centre of gravity is located much further away from the lower link ends. The moment of the implement weight counteracts the effect of the draught moment to the extent that top link sensing becomes imprecise. In any case, tractors can now pull longer implements than they can lift because it is not possible to add enough ballast to the front of the tractor to counterbal-

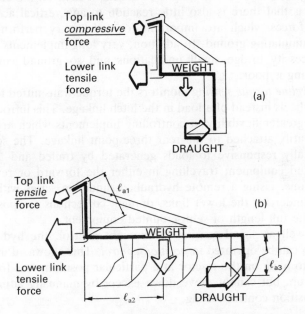

Fig. 8.12 The effect of the cross-shaft moments, due to the implement weight and draught, on the forces in the three-point linkage (for the moment arms ℓ_{a1}, ℓ_{a2}, ℓ_{a3}) causing: (a) a compressive force in the top link with a short implement; (b) a tensile force in the top link with a long implement.

ance the weight of the long implement. Some of the implement weight must be carried on a tail wheel which provides lift in conjunction with the lower links. Thus, the top link is made redundant for *semi-mounted* equipment, provided that the draught control signal is sensed elsewhere.

Bottom link sensing provides a control signal which is a truer measure of implement draught but adds to the design complexity because the loading in both lower links must be sensed as a sum, the proportion of the total load in each link varying considerably. The control signal is, however, less sensitive to a change in the horizontal draught of the implement than that with top link sensing. As the variation in draught is magnified in relation to the larger size of equipment being used at higher speeds, the reduction in sensitivity does not significantly affect the response of the draught control system. An adverse consequence of eliminating the effect of the weight of the implement on the control signal is that there is also little reaction to any vertical acceleration forces which are important for satisfactory performance over undulating ground. In addition, very long implements must of necessity bridge small undulations and so ground contour following is poor.

Driveline torque sensing monitors the torque transmitted to the rear wheels instead of a load in the hitch linkage. This introduces much greater flexibility in controlling implements which are not necessarily attached to the rear three-point linkage. The sensor is equally responsive to loads generated by trailed and front-mounted equipment travelling in either the forward or reverse directions. Using a remote hydraulic cylinder in series with the lift cylinder for the lower links, draught corrections are possible over the full length of semi-mounted equipment.

As well as signal sensing for draught control, the hydraulic system can be operated in *position control*. This allows an implement to hang on the tractor at a particular position; it is free to move up, but cannot move down except by manually adjusting the position control setting.

8.7.3 Front mounted hitches

Front-mounted hitches are a relatively recent development aimed at increasing the output of four-wheel tractors. Instead of having

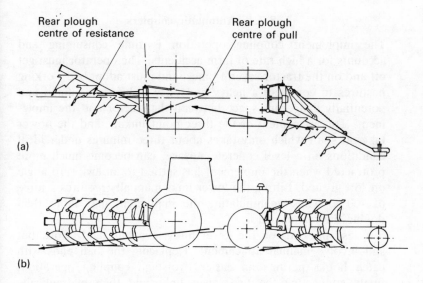

Rear plough
centre of resistance

Rear plough
centre of pull

(a)

(b)

Fig. 8.13 The effect of line of pull with front and rear mounted ploughs on (a) steering; (b) weight transfer.

to add ballast to the front of the tractor, the weight of a front-mounted machine can equally well provide a uniform axle loading for optimum traction *and* complete a field operation at the same time. Greatest popularity has been achieved by the combination of front- and rear-mounted ploughs, the 'push-pull' system (Fig. 8.13). Ploughing performance is improved by increasing the number of bodies through better traction. Individual front and rear ploughs are shorter and give better depth control on undulating ground compared with a single long unit.

The horizontal thrust from the front plough is, however, considerably offset from the tractor centre line and produces a large steering force which must be counteracted by adjusting the hitch of the rear-mounted plough to compensate. It also means that the front-mounted unit must be rigidly restrained on the linkage arms for directional stability and draught control is not available. Thus, front mounting is more suitable for equipment working directly ahead of the tractor and hanging in a fixed location which can be altered by manually adjusting the position of double-acting remote cylinders.

8.7.4   Automatic couplers

The implement coupling operation is time consuming and accounts for a high rate of farm accidents. The operator must get off and on the tractor several times, and must adopt bad working postures to complete a heavy and dirty task in a confined and potentially dangerous space between the tractor and the implement. The attachment of the three-point linkage and the power take-off shaft which only takes about three minutes under ideal conditions on a level concrete surface, can become much more protracted when the implement has settled to an awkward angle on soft ground. Lifting the lower links manually requires a force of 700 N, whilst manipulating the implement may more than double the manual effort.

Telescopic ends on the lower links save time and effort, but still involve manual attachment. Replacing the ball ends with catch hooks permits an easier two-stage coupling operation, firstly connecting the two lower links and then hydraulically adjusting the length of the top link. The hooks automatically lock onto the ball bushes which are fitted to the link pins on existing implements. Although there is no need to leave the tractor seat, it is not the easiest position from which to support and guide the top link during implement attachment. Unlatching the hooks is cord operated and requires a pull of up to 150 N, again in a twisted posture looking backwards.

An A-frame automatic coupler provides a quicker and simpler method of implement attachment, even when the misalignment between the tractor and implement is quite large (Fig. 8.14). It does involve a more expensive device, comprising an A-frame on the tractor linkage which mates with a similar frame on *each* implement.

Automatic couplers are also available for the power take-off driveline and are particularly useful with close coupled machinery, such as fully mounted, power-driven cultivators. The coupling connection is located in the telescopic part of the drive-line. The driveline section attached to the tractor power take-off shaft has a conical end which is centralised for docking by three tension springs to the three-point linkage so that the height of the docking cone can be adjusted through the linkage position. The driveline section attached to the machine has a tapered

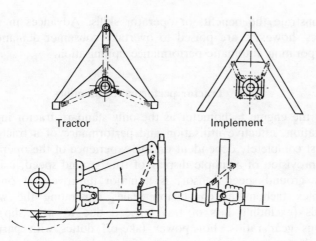

Fig. 8.14 Automatic coupler for attaching the power take-off shaft and the three-point linkage to mounted implements (after Kofoed, 1984).

spigot again suspended on two tension springs. Once the docking cone is level with the spigot, reversing the tractor completes the drive engagement before the links are attached.

8.8 INSTRUMENTATION AND AUTOMATION

All tractor cabs are equal to the extent that they must withstand the impact of the tractor rolling over and must be sufficiently sound-proof to keep the engine noise level at the driver's ear below 90 dBA. In addition, however, cab layout and control position are major sales features; for example, the welcome introduction of the flat uncluttered floor. Power steering, power brakes and the differential lock all provided substantial benefits for the operator without any apparent complication within the cab. It is surprising, therefore, to find that access, visibility and seating on new models continue to attract well-founded critical comment.

On the other hand, there has been little market pressure to encourage any major change to the basic tractor instrumentation and control systems. The low consumer priority for instrumentation and for automation is perhaps partly due to the psychological importance of being able to retain the challenge, and demonstrate the benefit, of operator skills. Advances in electronics, however, are poised to overtake consumer demand by incorporating automatic performance optimisation.

8.8.1 Tractor performance monitors

With the engine tachometer as the only standard tractor instrumentation, effective utilisation and performance of a tractor is almost completely dependent on the experience of the operator. The provision of a simple display of true ground speed, using a radar ground speed sensor, is a major improvement on the complex tachometer conversion scales illustrating the wheel speeds (excluding loss of travel speed due to wheelslip) for various gear ratios. For power take-off duties with partially loaded tractors, minimum specific fuel consumption and maximum work rates are achieved by increasing travel speeds because most tractors are designed so that the standard power take-off speed is obtained at or close to the rated engine speed. The direct display of travel speed, therefore, may provide a stimulus to the operator to maintain a higher level of efficiency.

Few farmers monitor fuel use, although the operator is aware of fuel consumption for various operations in terms of field area or operating time between tank refills. As the diesel fuel injection pump precisely meters the fuel to the engine, it is relatively simple to provide a highly accurate fuel flow meter. Encouraging fuel economy, however, is not the appropriate management strategy because operating the tractor at any point less than full power introduces a rate of work penalty. For a tractor of medium power, the extra labour cost consequential to the slower work rate virtually eliminates the saving in fuel cost, even discounting any crop yield loss from slower working or annual charges for larger equipment to complete the work at a slower rate but within the available time. The operator's simple fiat for minimum specific fuel consumption is: 'Shift up and throttle back' because an engine performs most efficiently in the upper part of the torque speed envelope, as shown by the specific fuel consumption map (Fig. 8.15).

A number of manufacturers now offer more comprehensive tractor performance monitors as an optional extra (or as standard on very large tractors). These monitors incorporate a radar

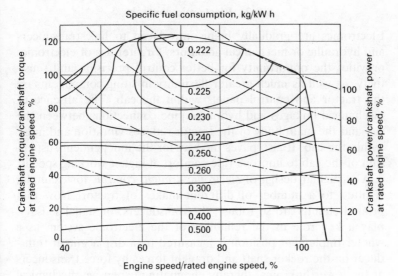

Fig. 8.15 Specific fuel consumption map (from NIAE, 1983).

ground speed sensor, an engine speed sensor and a wheel speed sensor in order to compute and display both true ground speed and wheelslip. By means of an implement status switch which indicates when the implement is lowered into work, a time clock and user entered machine width, the area covered and the overall rate of work can be displayed. Whilst the importance of wheelslip in draught work is undeniable, operator interpretation of the display is no less subjective than the occasional glance at the drive wheels. Without a measure of drawbar power to compare with the engine power, the operator does not have an effective yardstick for optimising wheelslip on different soils, except his own judgement.

It is quite feasible also to monitor drawbar power using existing sensors for driveline torque or by adopting an electronic draught sensing system (see section 8.8.2). With the inclusion of these additional data, prototype control systems are under active development for the complete automation of tractor engine performance and gear ratio selection. This will maximise rates of work as well as redirecting operational time from the driving function to more rewarding functions, such as monitoring machine processes and reacting to random failures.

8.8.2 Electronic controls

Electronics are gradually being introduced to link transducers and hydraulic or mechanical actuators. Greater use of electronics provides the opportunity for better control precision and sensitivity, as well as much greater flexibility in siting components on the tractor and control panels within the cab. The absence of mechanical linkages and hydraulic pipe connections between the cab and the main body of the tractor reduces vibration and noise within the cab. It also paves the way for the independent suspension of the cab to improve ride comfort at higher working speeds.

A particular design of *electronic draught control*, for example, is fitted to a number of different makes of tractor. Now that printed circuits are so robust, the transducers, or signal sensors, play a key role in the reliability of the electronic system as a whole. Implement position is measured by a displacement transducer on the rocker shaft and draught forces by force transducers at the lower link pins, thereby eliminating expensive mechanical spring assemblies which are always subject to hysteresis, or lag (Fig. 8.16). Both the displacement and force signals are mixed in a variable ratio which is selected on the control panel, the intermix ranging from 100 per cent draught to 100 per cent position control. The transducer signals are compared with the operator's nominal input setting on the depth control dial (Fig. 8.16). If the deviation exceeds the dead band, a threshold element and an amplifier output element activate the solenoid operated hydraulic valve to adjust the position of the three-point linkage until the error is eliminated. Sensitivity of the system is varied by adjusting the width of the dead band within which there is no remedial movement of the linkage, provided that any error signals do not exceed the threshold value. Variable sensitivity is an expensive feature to incorporate in mechanical systems.

A fast lift switch can be used to lift and lower the linkage at the headland without altering the setting of the depth control dial. Unintentional lowering of the links during transport is protected by a transport lock switch which only allows lowering in the 'field' position. Altering the depth control dial whilst the ignition is switched off automatically activates a starting lock which blocks the control system until the dial is turned to match the actual position of the linkage. An especially useful addition

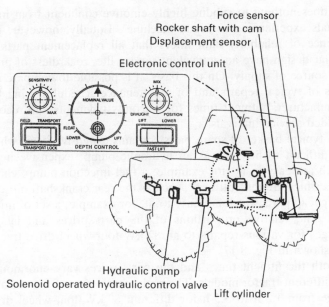

Force sensor
Rocker shaft with cam
Displacement sensor

Electronic control unit

Hydraulic pump
Solenoid operated hydraulic control valve
Lift cylinder

Fig. 8.16 Electronic draught control showing the location of the components on separate hydraulic (full line) and electronic (dotted line) control circuits, with the operator's control panel inset (adapted from Schrader, 1982).

is a remote push-button control to operate the linkage from the rear of the tractor when hitching implements.

8.9 REPLACEMENT PARTS AND FITMENT TIMES

Repair and maintenance costs depend on the reliability of individual machines, the prices of replacement parts, the labour charges associated with the standard fitment times and the maintenance schedules. Although component life is known to the manufacturer, every effort is taken to ensure that the information remains commercially confidential. Only the exceptional component failure receives wide publicity, and then often only for safety reasons to ensure that defective units are replaced by parts of modified design. The paucity of information virtually excludes reliability as an objective comparator in the choice of machines but does nothing to subdue highly emotive comment from individuals experiencing a 'rogue machine'. Equally, however, the absence of reliability data infers that all replacement parts of similar design have an equivalent service life, regardless of price and source of supply. On this basis, it is possible to compare the costs of typical repairs and of maintenance schedules using the manufacturers' fitment times for warranty payments and recommended service intervals.

A typical basket of spare parts contains expensive items which are difficult to fit, such as an hydraulic pump; expensive items which are easy to fit, for example, a fuel injection pump; cheap parts which are difficult to fit, such as the rear crankshaft oil seal; and cheap parts which are easy to fit, for example, a set of injectors. The relative proportions of the parts prices and labour charges for various repairs to an 82 kW four-wheel drive tractor are shown in Fig. 8.17.

Both the fitment times and the parts prices vary enormously for different tractor models. The replacement times for a radiator range from 1.3 h to 3.5 h for different 80 kW four-wheel drive tractors; and there is more than a three-fold variation between

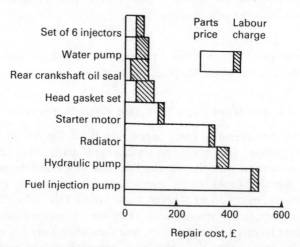

Fig. 8.17 The relative proportions of the parts prices and fitment charge for various repairs to a Massey Ferguson 2645 four-wheel drive tractor (after Hagger, 1986).

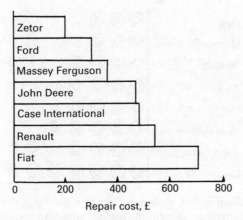

Fig. 8.18 The repair cost (parts price and labour charge) for replacing a radiator on different makes of four-wheel drive tractors of about 80 kW (after Hagger, 1986).

the cheapest and dearest total repair cost, taking the labour plus overheads charge at £15/h (Fig. 8.18). Even with a plethora of these bar charts, the sum of the limited range of repair costs examined for each tractor is not necessarily representative of the accumulated repair cost of component failures with use.

In order to demonstrate the underlying trend, the repair cost for replacing each part on each tractor can be ranked in order of expense – one point for the lowest cost and eight points for the highest cost in each case. Both a two-wheel drive model of around 60 kW and a four-wheel drive model of about 80 kW are included for each tractor make. For the repair cost ranking appraisal, the common parts on all tractors are head gasket set, fan belt, injectors, fuel injection pump, starter motor, alternator, clutch, and hydraulic pump; with three additional parts on the two-wheel drive models, namely, the rear crankshaft oil seal, front wheel bearing, and stub axle; and one extra part on the four-wheel drive models, namely, the front axle oil seal. This ranking procedure clearly identifies the manufacturers whose marketing strategy includes cheaper spares (Fig. 8.19).

Careful scrutiny of the maintenance schedules also reveals a large variation in both the service frequency and the oil requirements for different tractor models of similar power. This causes

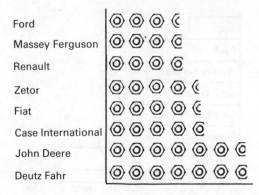

Fig. 8.19 Comparative ranking of repair costs for different makes of tractor, the number of nuts increasing with expense (after Hagger, 1986).

a two-fold difference in the annual cost of materials. Even though the variation in servicing time is absorbed as a farm overhead, the annual cost of materials for servicing a two-wheel drive tractor in the 60 kW class varies from £80 to £240 for 1000 hours of annual use.

8.10 FORECASTING TRACTOR DESIGN FEATURES

Agricultural tractors are multi-attribute products, with many features offering more than one benefit to the owner. This makes it difficult to establish the weighted value of each feature contributing to the worth of the machine for marketing purposes. Marketing products by trial and error to ascertain those features desired by the purchaser and to determine the choice between conflicting criteria is a very slow and costly process.

The alternative is to employ a quantitative approach to forecast the potential product value by adding a new feature. *Conjoint analysis* requires only rank-ordered input of two or more product features to statistically analyse their combined effect on the scaling of a dependent variable, such as the intent to buy. The basic premise of conjoint analysis is that any product or service can be viewed as a collection of functional, structural, social, psychological and/or economic attributes. In the appraisal of optional machine designs, these attributes represent buyer preferences for product features or benefits.

The potential of conjoint analysis may be demonstrated by a design appraisal of 12 features, each containing up to ten options or preference levels, for a 225 kW four-wheel drive tractor (Table 8.5). Firstly, each feature, containing more than two options, is examined in isolation from other features by over 300 individual buyers, their collated responses being used to rearrange the options in order of preference. The development of a fractional factorial design results in 26 *entirely* different combinations of feature options, or product feature packages, including four control feature packages. In the second stage, each member of the sample group sorts the product feature packages into one of four categories: an exceptional alternative, a desirable alternative, a questionable alternative, and an undesirable alternative. A rank order for all the product feature packages is completed by further sorting within each category. The conjoint algorithm then includes two data vectors, one vector being the preference levels of the options for each feature individually, and the other vector being the rank order of the product feature packages.

The option preference levels and the relative importance of the 12 features are shown in Fig. 8.20. The sum of the relative importance values for all the features is 100 per cent. Price, method of payment and warranty are all identified as factors of high importance, with price far outweighing the other two. Of the technical features, the power take-off options and the tyre options exert most influence on the purchase decision, whilst the relative importance of wheel adjustment is surprisingly high. Some features, such as the field lights, bonnet and hitch options, not unexpectedly rated low relative importance. From the study, the best possible feature package for a four-wheel drive tractor was, in order of relative importance:

1. price of $66 431;
2. direct drive power take-off;
3. 20.8 × 34 dual wheels;
4. one year full warranty with two-year engine and drive train warranty;
5. hydraulic wheel adjustment;
6. seven field speed gear ratios;
7. a straight annual finance contract;

Table 8.5 List of possible tractor design features and options

Feure	Option
1. Power take-off	(a) No p.t.o.
	(b) Hydrostatic p.t.o.
	(c) Direct drive p.t.o.
2. Hitch	(a) No hitch
	(b) Three-point hitch
	(c) Quick coupler
3. Cab	(a) No cab, rops only
	(b) Cab
	(c) Air-conditioned cab
4. Tyre	(a) 18.4 × 34 singles
	(b) 18.4 × 34 duals
	(c) 18.4 × 38 singles
	(d) 18.4 × 38 duals
	(e) 20.8 × 34 singles
	(f) 20.8 × 34 duals
	(g) 20.8 × 38 singles
	(h) 20.8 × 38 duals
	(i) 23.1 × 30 singles
	(j) 24.5 × 32 singles
	(k) 24.5 × 32 duals
	(l) 30.5 × 32 duals
5. Hydraulic remotes	(a) 2
	(b) 3
	(c) 4
	(d) 5
	(e) 6
6. Field lights	(a) 3
	(b) 4
	(c) 5
	(d) 6
	(e) 7
	(f) 8
7. Warranty	(a) 1 year – full
	(b) 2 year – engine & major drives
	(c) 2 year – full
	(d) 1 year – full; 2nd year engine plus major drives
8. Purchase method	(a) cash only
	(b) 3 year lease
	(c) 5 year lease
	(d) 7 year lease
	(e) straight annual finance contract
9. Wheel adjustment	(a) fixed
	(b) manual – move clamps and wheels
	(c) manual – sliding on axle
	(d) hydraulic

Table 8.5 (*Cont'd*)

Feature	Option
10. Bonnet	(a) tilting (b) fixed
11. Transmission ratios	(a) 4 field ratios (b) 5 field ratios (c) 6 field ratios (d) 7 field ratios (e) 8 field ratios (f) 9 field ratios (g) 10 field ratios
12. Price (not all combinations listed)	(a) \$51 000 (b) \$52 300 (c) \$53 800 (d) \$53 900 (e) \$54 700 (f) \$54 900 (g) \$77 143

(*Source*: Shoup, 1983)

8.   four hydraulic remotes;
9.   cab with air conditioning;
10. tilting bonnet;
11. hitch with coupler;
12. eight field lights.

8.11 SUMMARY

Farmers' machinery purchase decisions are strongly influenced by technical characteristics, after-sales service, negotiation, and bargaining.

There is a general tendency to buy farm supplies from the nearest source for convenience, both dealer and brand loyalty being quite high.

The acceptance of innovation is described by the adoption process which involves four main stages: awareness, interest, action, and confirmation. The length of the adoption process varies substantially amongst new products and practices. Simple ideas, easily tested, compatible with previous experience and yielding measurable results are likely to be adopted faster than

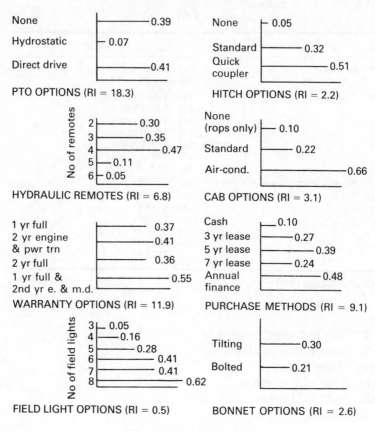

None ———— 0.39
Hydrostatic — 0.07
Direct drive ———— 0.41
PTO OPTIONS (RI = 18.3)

None — 0.05
Standard ———— 0.32
Quick coupler ———— 0.51
HITCH OPTIONS (RI = 2.2)

No of remotes
2 ———— 0.30
3 ———— 0.35
4 ———— 0.47
5 — 0.11
6 — 0.05
HYDRAULIC REMOTES (RI = 6.8)

None (rops only) — 0.10
Standard ———— 0.22
Air-cond. ———— 0.66
CAB OPTIONS (RI = 3.1)

1 yr full ———— 0.37
2 yr engine & pwr trn ———— 0.41
2 yr full ———— 0.36
1 yr full & 2nd yr e. & m.d. ———— 0.55
WARRANTY OPTIONS (RI = 11.9)

Cash — 0.10
3 yr lease ———— 0.27
5 yr lease ———— 0.39
7 yr lease ———— 0.24
Annual finance ———— 0.48
PURCHASE METHODS (RI = 9.1)

No of field lights
3 — 0.05
4 — 0.16
5 ———— 0.28
6 ———— 0.41
7 ———— 0.41
8 ———— 0.62
FIELD LIGHT OPTIONS (RI = 0.5)

Tilting ———— 0.30
Bolted ———— 0.21
BONNET OPTIONS (RI = 2.6)

(RI = relative importance)

(a)

their more complex counterparts, regardless of the cost-benefits involved.

Innovators, the personality group leading the adoption process, have more education, greater formal participation in organisations, higher social status, younger age, and a greater propensity for reading than the other personality groups.

Although farmers tend to be risk averters and conservative in their purchasing, they use few formal methods for investment appraisal.

The key technical feature in tractor choice is the power rating,

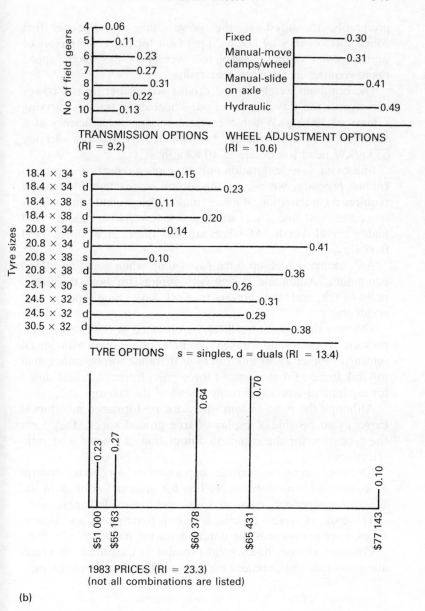

Fig. 8.20 The relative importance rankings of tractor design features and their options (after Shoup, 1983).

preferably measured at the power take-off. Specific fuel consumption varies by about 20 per cent for different tractors of similar power output. Complete coverage of the tractor speed range requires at least ten gear ratios.

The optimum weight on the driving wheels per unit of power varies with the travel speed. Four-wheel drive tractors carrying a mass of 100 kg/kW achieve maximum tractive efficiency at a speed of 6.5 km/h, whereas two-wheel drive tractors carrying 65 kg/kW need to operate at 10 Kkm/h.

Improving vehicle flotation only requires a reduction in ground contact pressure, but minimising subsoil compaction additionally requires a combination of lower total vehicle weight, dual wheels spaced at least one wheel width apart, fewer wheel passes, and higher travel speeds. All-wheel steering offers great operational flexibility.

An automatic pick-up hitch saves time when coupling trailed equipment. Automatic implement couplers for both the three-point hitch and the power take-off save time and reduce accidents.

The rear three-point hitch can be operated in position control, pressure control and draught control, the latter with signal sensing of either lower link forces or driveline torque rather than top link forces. Front-mounted three-point hitches are best suited for equipment working directly ahead of the tractor.

Although the main benefit of tractor performance monitors is currently to provide a display of true ground speed, they form the precursor for the complete automation of tractor gear ratio selection.

Electronic controls provide opportunities for better control precision and sensitivity, as well as for greater flexibility in the siting of components compared with mechanical linkages.

The costs of typical repairs, based on parts prices and fitment times, vary enormously for different tractor models.

Conjoint analysis holds great promise in quantitatively evaluating consumer preferences for new product feature packages.

9

CONTAINING FIXED COSTS

APOSTROPHE – SOUTH-SEEKING CHARIOT

Although public awareness of the 'differential gear' has been established through the importance of the device to the operation of all road vehicles, the mechanism has been in use for thousands of years. The invention of the differential gear first appeared in the design of a South-seeking chariot in China during the Han dynasty in the second century A.D.

The principle of its operation is shown in the accompanying diagram. Pivoting on the inside wheel, it requires two revolutions of the outer wheel for the chariot to negotiate a 360 degree turn if the spacing between the chariot wheels is the same as their diameters. It also requires two revolutions of the vertical gears in the differential mechanism to turn the figure through 360 degrees when the vertical gears are half the diameter of horizontal gears above and below them. Provided the drive train from each wheel to the differential unit is identical, the two horizontal gears rotate at the same speed but in oppo-

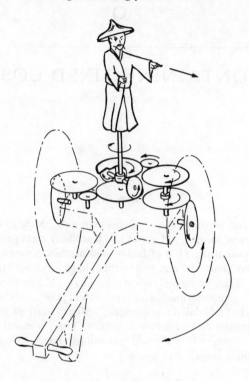

site directions when the chariot is pulled in a straight line. The vertical gears turn on the spot and the figure does not turn. As the chariot turns a corner, the inside wheel goes slower than the outer wheel so that the vertical gears not only rotate but also 'walk' round the horizontal gears in proportion to the difference in the wheel speeds which is linked to the extent of the corner.

By means of this simple design, the figure always points in the same direction whichever way the chariot is turned on a smooth surface. Uneven wheelslip, of course, gradually changes the direction but this can be corrected at intervals with reference to the sun. Hence the name: 'South-seeking chariot'. In business, as on any journey, it is important to be able to maintain direction, despite having to circumnavigate obstacles on the way.

(Source: Burstall 1963)

9.1 INTRODUCTION

Over the past decade production costs have increased more rapidly than farm output. This cost/price squeeze has led to a decline in farm profitability. As prices are unlikely to increase in real terms during a period of commodity surpluses, improvements in net farm incomes depend on reducing production costs. For management purposes, the costs of production are divided into variable costs which cover the outlay on seed, fertiliser and feed for specific enterprises, and overhead – or fixed – costs. Although it is the variable costs that are rising, there is little scope for savings because lower inputs reduce crop and animal production which, in turn, decrease farm output. For this reason, management emphasis is placed on containing fixed costs.

The major fixed costs are labour, power and machinery, and interest on borrowed capital – **Men, Machines and Money**. The term 'fixed costs' is a singularly unfortunate one because it implies that they cannot be altered. This is not so, but any change certainly involves drastic measures which affect the whole farm business. Improvements can only be brought about slowly over a period of time.

Arable farming with spring and autumn peaks tends to have a very uneven annual requirement for labour and machinery. Sadly for the stability of the rural communities, further reductions in regular staff are still possible by changing the enterprise mix, or by employing casual labour. Excessive depreciation costs can be lowered by replacing machinery less frequently but the importance of machine reliability must not be neglected. Interest charges can also become a crippling and potentially fatal burden on a business unless the level of borrowing is held in check by careful choice of the sources of finance and by taking advantage of contractors as an alternative to machinery ownership.

9.2 GUIDE TO FIXED COSTS

Items of farm expenditure are divided into *fixed costs* and *variable costs* (Table 9.1).
- **Fixed costs** are taken as business overheads which do not alter with small changes in output.

In contrast to fixed costs, the variable costs are readily allocated

Table 9.1 Items of farm expenditure divided into fixed costs and variable costs

Fixed costs	Variable costs
Regular labour (inc. farmer and wife manual work)	Seeds
Power and machinery running costs (except contract hire)	Fertilisers
Machinery and building depreciation	Sprays
Interest charges	Casual labour
Rent and/or landowning expenses	Contract hire
Rates	Feed
Miscellaneous office expenses, fencing, etc.	Miscellaneous, e.g. baler twine veterinary fees, etc.

to individual enterprises and vary in proportion to output.

On arable farms in the East of Scotland, fixed costs represent about two thirds of the total farm expenditure. As a large proportion of the total interest burden is carried by a small percentage of farmers, average interest charges would be misleading so that only imputed rent to cover mortgage interest payments for owner/occupiers is included in these fixed costs. For the period from 1980 to 1984, average fixed costs at constant 1984 prices were £420/ha for specialist cereal farms compared with £560/ha for general cropping farms (Fig. 9.1). The key for the diagram shows the relation between output, variable costs, *gross margins*, fixed costs, and *management and investment income*.

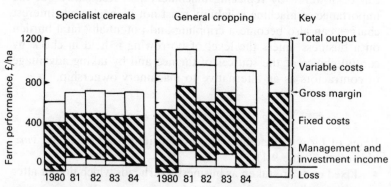

Fig. 9.1 Trends in farm performance at constant 1984 prices (adapted from Anderson, 1986).

- The **gross margin** is the total output less the variable costs and is applicable to financial data for individual enterprises as well as for the whole farm.

In calculating the management and investment income, any interest charges, ownership expenses and paid management have been excluded from the fixed costs, whereas imputed charges for the farmer's manual labour and rental value (where an owner/occupier) have been included.

- **Management and investment income** represents both the reward to management and the return on tenant's capital, whether borrowed or not.

Hence, deducting the interest charges and ownership expenses from the management and investment income gives the net profit (or loss).

Closer examination of the fixed costs reveals that by far the largest single item is power and machinery, with depreciation accounting for about half of the item and both repairs and fuel accounting for about a quarter each (Table 9.2). Often, however, labour is considered jointly with machinery because of their

Table 9.2 The relative importance of fixed cost items (excluding interest) for 1984–85 on two types of arable farms in Scotland

Fixed costs	Relative importance, % of total	
	Specialist cereal farm	General cropping farm
Labour (inc. farmer and wife manual work)	21	24
Machinery : repairs	10	10
depreciation	22	22
fuel and electricity	10	9
LABOUR AND MACHINERY COSTS	63	65
Property repairs	4	4
Rent and rates	25	24
Miscellaneous	8	7
OTHER OVERHEAD COSTS	37	35
TOTAL	100	100

(*Source*: SAC, 1986)

interaction, one substituting for the other, and together they represent about 65 per cent of fixed costs.

This labour and machinery complex is one of the most intangible areas within the farm business because of the wide range of cropping and stocking systems and the variety of combinations of men, machines and work methods. There is also little to assist by way of farm records. Recording labour usage, even where it is carried out, is a trap for the unwary. Records of most importance relate to operations carried out in peak periods, yet it is at this time that inaccuracies are most likely to occur because of the pressure of events. Task times tend to include all the hours worked, whether or not they are essential to the operation. These task times also vary with the machinery complement available. In view of these problems, the general level of efficiency in the use of labour and machinery is commonly investigated in financial terms by means of business indicators such as:

o **labour and machinery costs per hectare;**
o **total output per £100 of labour costs;**
o **total output per £100 of machinery costs;**
o **total output per £100 of labour and machinery costs.**

Comparison of the values of these business indicators for an individual farm with average farm management data establishes whether any further more detailed investigation is required (Table 9.3). Although these business indicators are output related, it must be emphasised that total output is important in its own right. The highly profitable farms seldom have less than average labour and machinery costs per hectare but the extra financial outlay produces much higher than average total output.

Table 9.3 Average levels of efficiency in the use of labour and machinery on two types of arable farms in the East of Scotland for three crop years, 1982–84

Business indicator	Specialist cereal farms	General cropping farms
Total labour cost per hectare, £/ha	121	182
Total machinery cost per hectare, £/ha	166	232
Total output cost per £100 of labour costs, £	541	512
Total output cost per £100 of machinery costs, £	392	403
Total output per £100 of labour and machinery, £	227	225

9.3 FARM ORGANISATION AND PLANNING

A major policy change within the farming system often has greater repercussions on the business than was originally envisaged. Accurate forward planning involves the application of formal procedures to study the work capacity of a whole team of men and machines performing a series of operations over an extended period of time. These planning procedures include:

○ **the gang workday chart;**
○ **farm work scheduling;**
○ **arable farm planning models.**

The more complex the planning procedure, the less widely they are used because of the mental tenacity involved in distilling the optimum solution from a large number of possibilities. Whilst the laboriousness of the manual methods can be largely eliminated by linear programmes, planning models suffer from the paucity of system performance data available. Much of the planning time is used in judging the most appropriate choice of inputs; the more skilful the selection, the better the final work schedule. The results of this type of arable farm planning exercise, at the very least, provide reassurance that the system capacity is adequate and may well allow a potential imbalance of resources to be rectified before the policy is adopted.

9.3.1 Gang workday chart

The gang workday chart is a particular application of the principle of resource allocation described in section 2.4.2. The chart is built up from a series of blocks which show the number of workers and the duration of individual operations throughout the year or for a peak period of the year (Fig. 9.2).

The number of workers is identified on the vertical axis of the chart and the number of days available for fieldwork in each month is indicated on the horizontal axis. The workdays can be reduced to allow a greater safety margin on difficult soils by reference to the soil workday data (see section 6.3). For planning purposes, each workday is taken as eight hours, and overtime is held in reserve to overcome the inevitable problems encountered in a difficult season.

The data for the construction of the chart comprises:

○ **number of workers;**

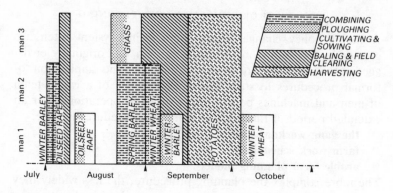

Fig. 9.2 The gang workday chart showing the autumn labour peak for a general cropping farm of 120 ha with potatoes.

○ **number of field workdays required**
  (area of work, ha, and rate of work, ha/h);
○ **operational window**
  (earliest start date and latest finish data).

Task times can be altered by varying the rate of work appropriate to different sizes of machine and by adjusting the area of work with respect to the enterprise mix. Alternative enterprises can change the operational window; for example, replacing spring barley by winter barley. Judicious rearrangement of these resources can result in a more even labour requirement.

The autumn labour peak, shown in Fig. 9.2 for a general cropping farm of 120 ha, is dominated by a large block for harvesting the potato crop. One possibility is to rent the 16 ha of potato ground annually to a potato merchant, to increase the winter barley area from 8 ha to 32 ha and the oilseed rape from 8 ha to 12 ha in place of some of the spring barley, whilst the winter wheat and one year grass ley areas remain at 24 ha and 12 ha, respectively. The gang workday chart for the new cropping plan is shown in Fig. 9.3. Most of the tasks dovetail together well for a team of two men instead of three, except for harvesting winter wheat. Rather than employ a third man, a contractor could be used to overcome the remaining labour peak (see section 9.6.1). It is noticeable, however, that smoothing the labour requirement

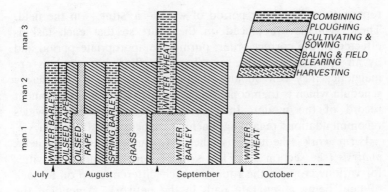

Fig. 9.3 The gang workday chart for a general cropping farm when potatoes are excluded from the rotation, the winter wheat harvesting being contracted out to save one man.

has eliminated all the slack periods over a two-month period and this emphasises the importance of not only planning the critical operations but also taking account of such items as staff holidays.

This method is simple to use but suffers from several disadvantages. It may involve a number of attempts on different charts to satisfactorily solve a problem. The levels of skills of the labour team are not differentiated, so there is no distinction between skilled and unskilled staff. Equally, it is difficult to show the detail of tasks running in parallel over an extended period and involving the whole team, such as potato harvesting in the morning followed by cereal harvesting later in the day. It must also be borne in mind that the technique only evaluates the technical feasibility of the various options and does not include any financial appraisal.

9.3.2 Farm work scheduling

Some of the limitations of the gang workday chart are overcome by *farm work scheduling*. One method of farm work scheduling involves the use of magnetic symbols to build up an analogue model on a steel-backed monthly time chart. Each symbol

represents a four hour period of work – a 'stint' – in the field.
These symbols are placed on the chart so that each task is
effected in the correct order, during the appropriate period and
by an operator with the suitable skills. Since the symbols are
magnetic, they can be moved about to achieve a satisfactory
schedule which is then copied onto a paper chart as a permanent
record of the model. Interpretation of this model provides
recommendations regarding machine capacity and farm policy.

Farm work scheduling is, therefore, an exercise in *network
analysis* (see section 2.4). The symbols represent farm activities
for various crops in the rotation, each different crop on a block
of land being a separate path in the network. Amending the
balance of the available resources changes the critical path
through the network. A case study proceeds in three stages, the
first being for farm data collection. Next follows the organisation
and arrangement of farm data, culminating in the construction
of the model of the work schedule.

Stage 1 Data collection

It is usual to start by defining the cropping plan over a two-year
period to cover a transition from one year's crops to the next,
for example in scheduling the workload during an autumn peak
(Table 9.4). Individual blocks of land in the crop rotation are

Table 9.4 Two-year farm cropping plan with letters identifying particular
blocks of land

	Crop area, ha					
Year 1 Year 2	Grass	Wheat	Barley	Rape	Etc.	Total year 2
Grass	50	—	50[B]	—		100
Wheat	50[E]	150[D]	—	100[A]		300
Barley	—	100[C]	—	—		100
Rape	—	—	—	—		—
Etc.						
Total year 1	100	250	50	100		500

(*Source*: Barrett, 1982)

identified by letters in the order in which the work is to be started, the absence of a letter indicating that no work is required. Some crop transitions present few organisational problems: for the 100 ha block A, there is adequate slack for most contingencies between harvesting the oilseed rape in late July and sowing winter wheat by mid October. Other crop transitions are more critical, with only a short timespan available for a sequence of operations: for the 100 ha block C, late harvesting of the wheat crop leaves little time for drilling the subsequent crop of winter barley by mid September.

Stage 2 Data organisation

For each block of land, a 'menu' of operations is selected to achieve the crop transition (Table 9.5). Work rates for particular machines and job allocations to particular staff are discussed and agreed. Using these work rates, the number of four hour stints needed to fulfil every activity for an individual block of land is entered on a work sheet (Table 9.6). The total number of stints for all the operations on all the blocks is a measure of the total workload.

Stage 3 Data analysis

Symbols bearing the activity code number, for example C10, are then arranged on the chart initially at two per day. This two stint day in the field probably means a 10-hour working day, with the inclusion of travelling and so on. Six such days gives a moderate working week. In order to allow for more overtime, a third symbol on alternate days gives a working week of about 75 hours – heavy pressure over prolonged periods. Sundays are always left free to allow a margin for contingencies but, in practice, advantage would be taken of favourable weather to complete urgent field work. Adverse weather is represented by reducing the number of calendar days available for field work.

Starting with the crop areas, the sequence of operations, the team size and the machinery complement, farm work scheduling can predict progress and critical activities. Once the chart has been constructed for a given set of information, however, it is very tedious making alterations. The iterative procedures

Table 9.5 Menu of operations and activity codes allocated to different blocks of land in the crop rotation

Crop rotation year 1	Crop harvested	Rape	Barley	Wheat	Wheat	Grass	
	Crop area, ha	100	50	100	150	50	
	Block of land	A	B	C	D	E	F
				ACTIVITY CODES				
HARVESTING							
10	*Spray, desiccant, etc.*	—	—	C10	—	—	
11	*Mow, Swath, Windrow*	A11	—	—	—		
12	*Harvest, forage, roots*	A12	B12	C12	D12		
13	*Transport, produce of 12*	A13	B13	C13	D13		
14	*Secondary Harvest, bale, save tops*	—	B14	—	—		
15	*Secondary Transport, produce of 14*	—	B15	—	—		
16	*Crop, Spread, gather debris*	A16	—	—	—		
17	*Burn*	A17	—	C17	D17		

		Wheat	Grass	Barley	Wheat	Wheat
TILLAGE						
20	*Spray*	—	—	—	—	E20
21	*Slurry/FYM*	—	—	—	—	—
22	*Break/Stubble*	—	—	—	—	E23
23	*Subsoil*	A23	—	—	—	E24
24	*Plough*	A24	—	—	—	E25
25	*Secondary Tillage*	A25	—	C25	D25	—
26	*Fertiliser*	—	B26	—	D26	E27
27	*Secondary Tillage*	A27	—	C27	D27	—
28	*Secondary Tillage or Nominate*	—	—	—	—	—
29	*Work Down*	A29	—	C29	D29	E29
SOWING						
30	*Drill, Plant*	A30	—	C30	D30	E30
31	*Roll*	—	—	C31	—	E31
32	*Harrow*	—	—	—	—	—
33	*Spray*	A33	—	—	D33	—
34						
	Crop sown	Wheat	Grass	Barley	Wheat	Wheat
	Crop rotation, year 2					

(*Source*: Barrett, 1982)

Table 9.6 Worksheet for recording the number of stints for every activity on an individual block of land

Activity code	Operation	Purpose or crop	Area, ha	Workrate ha/stints	Number of stints
C10	Spray	Pre-harvest desiccant	100	20	5
C12	Harvest	W wheat	100	5	20
C13	Transport	Grain			20
C17	Burn	Straw	100	10	10
C25	Disc × 2		200	12.5	16
C27	Springtine		100	10	10
C29	Harrow		100	12.5	8
C30	Drill	W barley	100	15	7
C31	Roll		100	25	4

(*Source*: Barrett, 1982)

required to optimise resources, including cash considerations, are best handled by computer.

9.3.3 Arable farm planning

Linear programming is a more complex technique which is often used to find an optimum cropping plan for a given team size and machinery complement. It can also be adapted to determine the maximum farm profitability by varying the cropping plan, the number of men and the sizes of the machines. A program comprises an '*objective function*' which expresses the objective of the analysis, for example the profit to be maximised, within a number of *constraints*, such as:

○ **available workdays;**
○ **operational timespans;**
○ **activity sequence;**
○ **land area;**
○ **crop sequence;**
○ **range of crop areas;**
○ **range of team sizes;**
○ **range of machine numbers.**

The nature of the information necessary for the analysis is similar to that for other comprehensive planning procedures, but a greater amount of detail is required to ensure continuity within the data sets for labour, machines, crops and rotations. The reason for this is that a linear program, as its name implies, assumes that inputs, outputs and the relationships between them are infinitely divisible. Although it is possible to deal in whole numbers by integer programming, the time needed to solve a problem increases enormously. It is generally more efficient, and also more informative, to examine the solutions without whole numbers and then, if necessary, to re-solve the model with the team size set to the whole number nearest to the optimum value and the machine sizes altered to avoid fractional complements.

Although the cost and availability of resources and the relative profitability of crops can alter dramatically in a fairly short period, the major imponderable factor in a farm planning exercise is the weather which influences the availability of workdays. The farmer's expectation of the occurrence of workdays usually reflects his attitude to risk. Rather than expecting average conditions to prevail, it is prudent to err on the side of caution by adopting a level which occurs for seven or eight years out of ten (see Chapter 6). By dividing the farming year into a number of fortnightly or monthly periods, a number of levels of workdays can be allocated for different operations during each period.

Conflicting demands by different operations for men and machines can be solved with the inclusion of penalties for untimely operations (see Chapter 5). Instead of restricting an operation to a given timespan, priority is arranged to minimise the penalty costs. The least delay is accorded to the most critical operation which, by definition, always incurs the greatest financial penalty through mistiming. More men and machines may reduce the penalty costs to the extent that overall profit is improved.

The farm profit is obtained by deducting the annual costs of men and machinery and other fixed costs for buildings and rent from the sum of the crop gross margins. The other fixed costs are independent of any changes to the crops, men and machines. Labour costs are quite straightforward and fuel costs can be related to tractor power and annual use (see section 4.4). Calculating the annual cost of machinery by a discounted cash flow method includes the effect of initial price, repair and maintenance costs, resale value, interest charges and inflation (see section 4.6).

The complete model is then capable of assessing different management strategies for individual arable farms and of selecting that combination of activities which maximises farm profitability within a range of farm conditions.

9.4 MACHINE REPLACEMENT

Machinery breakdowns are a major source of irritation at any time but may incur associated losses which are far in excess of the direct repair costs by delaying critical field operations. Consequently, the quest for machine reliability has a substantial influence on machinery replacement policy. As tractors become more powerful and machines become larger, the decline in fleet size increases the dependence on fewer units and the demand for greater reliability. The greater the complexity of the complete machine, however, the greater the chance of failure of an individual component, so that the price of improvements in reliability becomes progressively more expensive. As detailed reliability data on farm equipment is virtually unobtainable, machine replacement is largely based on economic pointers to minimise the holding cost of individual machines and to eliminate excessive fluctuations in machinery investment from year to year.

9.4.1 Probability of failure

It is only in the aircraft industry that every component must have a specified operational life which allows the part to be replaced *before* catastrophic failure in service. The automotive industry treats component failure rates as commercially confidential. Whilst consumer groups attempt to collate reliability data on domestic appliances and on cars from statistical analyses of user surveys, such information is often too late to assist the purchaser. The data in Table 9.7 are likely to comfort motorists already driving Vauxhall cars and further irritate the dissatisfied owners of Austin cars, but provide no security to potential customers now in the market for a subsequent model of either make. The manufacturer is quickly alerted to a serious design fault, either through accidents or through consumer complaints, long before

Table 9.7 A consumer survey of component failure
rates for three popular makes of car

Make of car	Component cost, £	Component failure rate, %		
		40 000 km	56 000 km	72 000 km
Battery				
Vauxhall Cavalier 1.6	43	1.5	3.9	5.8
Ford Sierra 1.6	25	16.6	22.8	29.7
Austin Montego 1.6	43	14.1	29.9	43.2
Alternator				
Vauxhall Cavalier 1.6	65	2.3	3.5	5.4
Ford Sierra 1.6	42	2.2	6.3	7.7
Austin Montego 1.6	101	15.2	17.9	25.4
Starter motor				
Vauxhall Cavalier 1.6	61	1.1	2.3	3.3
Ford Sierra 1.6	50	1.1	4.4	8.5
Austin Montego 1.6	132	5.4	14.1	26.1
Clutch				
Vauxhall Cavalier 1.6	79	0.4	14.0	27.0
Ford Sierra 1.6	94	14.0	30.0	62.0
Austin Montego 1.6	135	5.3	15.0	41.0

a random survey can amass adequate evidence. The existence of the original problem is only acknowledged retrospectively through Service Book amendments explaining the procedures for fitting the modification or by the introduction of a new model. There remains, therefore, a wide variety of unrelated failures which may lead to emotive but unsubstantiated outbursts of poor reliability relating to a particular model or, in their worst combination, to aspersions on the 'rogue' machine.

It is important that the failure rate of any machine component remains at a consistently low level over its useful life, after which wearing-out failures increase rapidly (Fig. 9.4). A slightly higher failure rate which is inevitable during the initial running-in period creates the familiar 'bath tub' shape of the failure rate curve with time.

The failure rate determines the *reliability* of the component.

- **Reliability** is the probability that a component or system of components will perform its specified function for a given period of time under the specified environment.

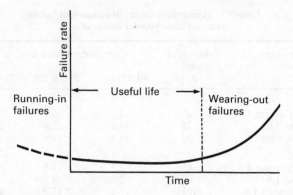

Fig. 9.4 The 'bath tub' shape of the component failure rate curve with time.

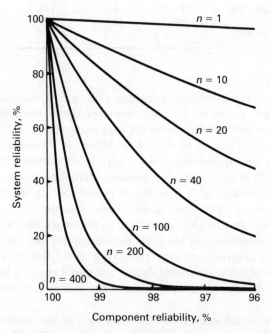

Fig. 9.5 The effect of series combinations of 'n' machine components upon the reliability of the complete system (adapted from Liljedahl *et al.*, 1979).

For a constant rate of failure per unit of time, the reliability of a component is:

$$r_c = 100/e^{(f \times t)} \qquad \ldots [9.1]$$

Reliability, % = 100/exp (Failure rate per unit time, 1/h × Time, h)

The reliability of the complete machine depends on the number and reliability of the individual components. For a mechanism depending on a series of 'n' individual parts, each with a different reliability, r_1 to r_n, the system reliability is:

$$R_s = 100 \, (r_1 \times r_2 \times r_3 \times \ldots \times r_n)/100^n \quad \ldots [9.2]$$

System reliability, % = Product of Component reliabilities, %

This equation can be simplified when all the components have the same reliability. The effect of increasing complexity of machinery, with more parts of similar reliabilities in series, is displayed in Fig. 9.5. Component reliability must meet very high standards to provide an acceptable system reliability for a machine with large numbers of dependent parts because failure of any one component incapacitates the machine.

9.4.2 Period of ownership

In the absence of detailed component reliability data, the vehicle replacement policy for fleet operators can be based on the accumulated repair costs for individual vehicles (Fig. 9.6). Any particular vehicle which has required a disproportionate amount of repairs should be replaced early, followed by an enquiry into the causes of the high rate of component failures. It is not unlikely that as much of the blame may be apportioned to the inadequacies in driving skills of the 'accident-prone driver' or to poor maintenance, as to excessive machine usage and to inherent production faults in the vehicle (see section 4.4.2)!

On most farms, however, there are seldom enough tractors or machines of the same type to provide an accurate comparison of repairs which, by their very nature, tend to be 'lumpy'. Some major overhauls and replacements can be anticipated with a fair degree of accuracy; for example, tyre wear is clearly visible and engine life is reasonably predictable. The high cost of drive tyres often initiates tractor replacement activity. After all, if tractor

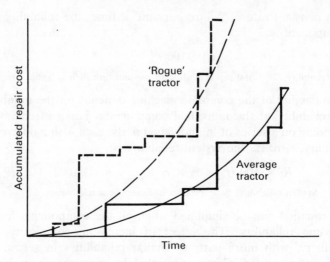

Fig. 9.6 Accumulated repair costs for a 'rogue' tractor compared with those for an average tractor.

replacement is under consideration, it is sensible either to trade-in the old tractor while there is still a little wear left on the current set of tyres or procrastinate long enough to get a reasonable return from the capital outlay on another set of tyres. Selling an old tractor with a brand new set of tyres is a buyer's bargain.

For forward planning purposes, the optimum time for replacing a machine is at the point when the annual ownership cost reaches a minimum. The annual machinery costs from actual cash flows for the full period of ownership incorporate decremental depreciation, accumulated repairs as a percentage of the purchase price, interest charges, insurance premiums based on the written-down value, and tax allowances all adjusted to current monetary values (see section 4.6). A separate cash flow analysis is completed for each different period of ownership. As a machine gets older, the annual depreciation charge declines but the annual repair cost increases. The variation in the annual costs for ownership periods from one to eight years is shown in Fig. 9.7 for a tractor with an initial price of £13 855. In this example, it is assumed that the inflation rate is 5 per cent, the investment rate is 8 per cent, the loan interest rate is 11 per cent and the annual use of the tractor is 1000 hours. If the present annual cost is £2900 for a period of ownership of two years, then this is the cost which is incurred in *each* of the two years. The annual ownership cost for the tractor rapidly declines as the ownership period is extended to three years but, thereafter, the annual cost curve slopes gently towards a minimum point and then the gradient slowly increases again.

The minimum point on the annual cost curve is located at its intersection with the marginal holding cost curve (Fig. 9.7). The marginal cost represents the extra cost incurred by keeping the machine for an additional year and, for example, replacing at the end of the fourth year instead of at the end of the third year, and so on. At the end of the first year, the marginal holding cost is equivalent to the present annual cost. Thereafter, the marginal holding cost for the additional year must be less than the average holding cost in order to pull down the present annual cost which is the average holding cost in each year over the complete period of ownership. As soon as the marginal holding cost exceeds the average holding cost, the present annual cost for the machinery must increase. Some of the marginal cost may well accrue in earlier years because the machine is purchased by means of a mortgage over the period of ownership: the longer the term for the mortgage, therefore, the higher the proportion of interest relative to capital repayment in the early life of the machine.

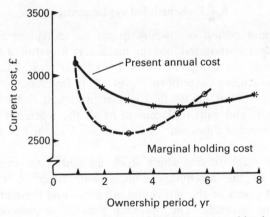

Fig. 9.7 The optimum period of machine ownership is when the marginal holding cost is equal to the present annual cost, using discounted cash flows.

Ideally, machines should be replaced once their age exceeds the period of ownership for which the annual cost is a minimum. The extra cost of not doing so can be obtained from the marginal cost curve. This extra cost is not as great as might be expected, but it is realistic. The higher average investment which is required for newer equipment is substituted by higher annual repair bills for older machines. This is more appropriate to the initial purchaser of any machine. For the next buyer, the cash outlay for repairs to older machines may be substantially reduced by salvaging parts from scrapyards and/or by completing the work in the farm workshop. The annual cost curve must then be revised, beginning with the second hand value, assuming the appropriate decremental depreciation values in accordance with the true machine age and accumulating repair costs at a slower rate which reflects the 'make do and mend' policy. These adjustments to the costs result in another annual cost curve, lower than the first one, with a minimum point which defines the optimum period of ownership for the second owner. Whilst this may reduce the capital investment and the annual machine costs, the poorer reliability of the older machine may lead to excessive downtime and incur severe yield penalties from untimely operations.

9.4.3 Scheduled replacement

The optimum period of ownership for an individual machine must also be incorporated into the machinery investment strategy for the machinery complement on a particular farm. The intention is to achieve a uniform level of re-investment in new machinery, without undue fluctuations in capital demand from year to year, and with replacement of all the machines rotated over a reasonable timespan. The *whole stock method* is used for this purpose.

The first stage of the whole stock method is to prepare an inventory of the machinery complement, together with the current list prices of the individual machines and their proposed replacement intervals. The optimum periods of ownership for different types of machinery vary with their rates of depreciation and their accumulated repair costs, some machines remaining serviceable for longer than others. Initially, it is convenient to adopt theoretical replacement frequencies of 3, 6 and 12 years. Over the 12-year timespan, some machines (for instance, combine harvesters) would be replaced four times, some (for example, tractors) twice, and others (such as trailers) only once.

Using depreciation rates appropriate to the type of machine and the replacement interval, the additional capital required to finance replacement can be calculated for each machine and as an annual total for the machinery complement over the complete timespan of 12 years. Whilst this timespan may be further ahead than most farmers would care to consider, it is much easier to make shorter term amendments once the broad framework has been established. The method is easy to apply and can be reviewed every few years.

As an illustration of the application of the whole stock method of investment appraisal, consider a 300 ha arable farm growing a range of crops all suitable for combine harvesting. A combine harvester, an 130 kW tractor, and a car are replaced at three-yearly intervals; three 65 kW tractors, a Land-Rover and some machinery are replaced at six-yearly intervals; and the remainder of the 22 items on the inventory are replaced every 12 years (Table 9.8). The replacement costs, as percentages of the current list prices of the various machines, are derived from the average falls in machinery values (see Table 4.3), but other depreciation rates can be substituted if considered desirable. This leads to an average machinery investment of just under £29 000 per annum, though much depends on which machines are replaced in which order.

The order of machinery replacement is built up on a chart by blocking in the most expensive inventory items first (items 1 and 6 in Fig. 9.8). With a certain amount of rearrangement of the cheaper items of machinery, the annual capital outlay can be adjusted to closely follow the chosen level of investment. Premature replacement of a particular machine, for whatever reason, will involve amending the replacement policy but the consequences are immediately apparent.

Reworking the same example with replacement frequencies of two, four and eight years, the annual replacement cost would rise to £34 800.

It is also possible to represent the effect of inflation on the chart by calculating the annual increase. At a constant inflation

Table 9.8 Replacement interval and cost for individual machines on a farm inventory

Item no.	Description	Current price, £	Depreciation, %	Replacement cost, £, at different intervals		
				3 yr	6 yr	12 yr
1.	Tractor, 130 kW, 4WD	38 000	49.5	18 810		
2.	Tractor, 65 kW, 4WD	18 500	72		13 320	
3.	Tractor, 65 kW, 2WD	15 000	72		10 800	
4.	Tractor, 65 kW, 2WD	15 000	72		10 800	
5.	Rough terrain forklift	16 000	90			14 400
6.	Combine harvester	40 000	49.5	19 800		
7.	Baler	5 000	72		3 960	
8.	Bale accumulator	1 000	72		720	
9.	Bale loader	1 000	72		720	
10.	Fertiliser spreader	4 000	72		2 880	
11.	Drill, 4m, seed only	7 000	90			6 300
12.	Sprayer, 12m, mounted	2 000	72		1 440	
13.	Trailer 1, 8t, grain	4 000	80			3 200
14.	Trailer 2, 8t, grain	4 000	80			3 200
15.	Plough, 6f, reversible	6 000	80			4 800
16.	Subsoiler	6 000	80			4 800
17.	Discs, 6.1 m	5 500	72		3 960	
18.	Cultivator, coil tine	3 000	80			2 400
19.	Harrows, chain	1 500	80			1 200
20.	Rolls, cambridge, 6.1 m	3 500	80			2 800
21.	Land Rover	10 000	72		7 200	
22.	Car	15 000	60	9 000		
Total for replacement interval				£47 610	£55 800	£43 100
				÷ 3	÷ 6	÷ 12
			Annual average	£15 870	£ 9 300	£ 3 592
			Annual total	£28 762		

(*Source:* Barrett, 1985)

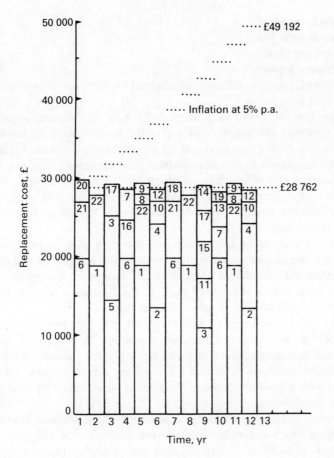

Fig. 9.8 A machinery replacement chart showing the order of replacement of individual machines which are identified by their inventory item number in Table 9.8 (from Barrett, 1985).

rate of 5 per cent, the replacement cost of £29 000 in the first year would increase to almost £50 000 by the twelfth year. This emphasises the need to set aside sufficient funds from revenue to ensure that the machinery stock is retained in good order.

9.5 SOURCES OF FINANCE

Machinery financing falls into four main categories:

○ **bank overdraft;**
○ **bank term loan;**
○ **hire purchase;**
○ **finance leasing.**

A particular source of finance is selected on the basis of *cost, cash flow and convenience*.

The *cost* of borrowing at a fixed rate of interest usually increases with the length of period of the loan because the lender demands a higher rate of return as a compensation for the higher risk of the funds being unavailable. It is, therefore, cheaper to finance machinery purchase through short-term borrowing, but it can lead to inconvenience because short term interest rates are more volatile and the availability of funds is subject to more frequent review.

With purchase or hire purchase, the *cash flow* is improved because the farmer becomes the owner of the equipment and can claim capital allowances. Any interest charges on overdrafts or loans to purchase machinery used in the business are a normal business expense and are allowable in full.

Ease and *convenience* rank very high for hire purchase and finance leasing arranged through machinery dealers. Low cost finance through packages of these types are occasionally offered by machinery manufacturers for limited periods to promote sales, often with a greater response than could be achieved by straight price discounting.

With a variety of deposits, repayment schedules and interest rates quoted for varying periods, the difference in the cost of funds is difficult to determine and the choice of the source of borrowing is often more dependent on the farmer's risk aversion.

9.5.1 Comparing the cost of borrowing

Sources of finance require careful scrutiny to translate them into a common format for comparative purposes. The simplest approach is to compare the cost of a bank term loan, hire purchase, and leasing with a bank overdraft. Indirectly, this then allows comparison between any of the four sources of finance.

The standard method of comparing interest rates is by the *annual percentage rate*.

- The **annual percentage rate** is the interest payable on £100 borrowed for a period of one year.

Whenever credit terms are offered in conjunction with a sale, such as a hire purchase transaction, the annual percentage rate must be quoted. Banks, building societies, hire and leasing companies are not covered by this legislation.

Banks and building societies refer to a *nominal rate*.

- The **nominal rate** is the true interest rate expressed on an annual basis.

Unlike the annual percentage rate, the nominal rate does not allow for the fact that the *real interest rate* is levied on the outstanding balance at intervals throughout the year – either monthly, quarterly or half-yearly.

- The **real interest rate** is the interest rate quoted for a given charging period.

The most familiar example of monthly compounding is the credit card loan or credit charge on merchants' invoices. A true interest rate of $1\frac{1}{2}$ per cent per month is equivalent to an annual percentage rate of over $19\frac{1}{2}$ per cent instead of a nominal rate of 18 per cent per annum (Table 9.9).

$$i_{apr}/100 = (1 + i_r/100)^n - 1 \qquad \ldots [9.3]$$

Annual percentage rate, % $\propto$ Real interest rate, %, compounded by the Number of part yearly intervals

Table 9.9 Annual percentage rates of interest compounded from the annual nominal rates

Annual nominal rate, %	Annual percentage rate, %		
	Monthly compounding (e.g. credit card)	Quarterly compounding (e.g. bank overdraft)	Half-yearly compounding (e.g. bank loan)
6	6.2	6.1	6.1
8	8.3	8.2	8.2
10	10.5	10.4	10.3
12	12.7	12.6	12.4
14	14.9	14.8	14.5
16	17.2	17.0	16.6
18	19.6	19.3	18.8
20	21.9	21.6	21.0
22	24.4	23.9	23.2

For a bank overdraft with interest at a nominal rate of 12 per cent charged quarterly, the real rate of interest is 3 per cent quarterly which, when compounded, gives an annual percentage rate of 12.55 per cent. Building societies charge interest on the outstanding balance at the *beginning* of the year, so that the annual percentage rate is higher than the nominal rate because some of the capital is repaid during the year.

The most confusing rates of interest, however, are the *flat rate* loans quoted for hire purchase and finance leasing.

- The **flat rate** is the interest rate levied on the initial sum, over the whole period of the loan.

The flat rate makes no allowance for the reduction in debt that occurs as the repayments are made. The interest is added to the amount borrowed, the total being repaid in equal instalments over the period of the loan.

Hire purchase usually entails an initial deposit with repayments monthly or quarterly in arrears. The real rate of interest may be calculated by discounted cash flow methods or by the following equation:

$$i_r = 2 \times n_i \times I \times 100 \, [L \, (N_i + 1) + I \, (N_i - 1)/3] \qquad \ldots \, [9.4]$$

Real interest rate, % $\propto$ Number of instalments per year, dim. × Total interest charge, £ ÷ [Amount lent, £ × (Total number of instalments, dim. + 1) + Total interest charge, £ × (Total number of instalments, dim. − 1)]

On a loan of £1000, with quarterly payments over three years, a flat rate of 6 per cent implies a total interest charge of £180. This gives a total hire purchase payment of £1180, or £98.33 per quarter. The real rate of interest is 2.6 per cent – an annual percentage rate of 11 per cent. The conversion from flat rates of interest to annual percentage rates is given in Table 9.10 for quarterly payments in arrears over one to four years. As just over half the original amount of borrowed capital is outstanding throughout the period of the loan, the annual percentage rate is obtained very approximately by doubling the flat rate and subtracting 1 per cent.

Although leasing contracts also use flat rates, the additional stipulation for rentals to be paid in advance increases the annual percentage rate. The ready reckoner in Fig. 9.9 transforms the quotations involving two-, three-, four- and five-year periods and monthly, quarterly and half-yearly rentals into the equivalent

Table 9.10 Quarterly hire purchase payments over one to four years and the conversion from flat rates of interest to annual percentages rates (APR) on a loan of £1000

Time period, month	12		24		36		48	
Flat rate of interest, %	Payment, £	APR, %	Payment, £	APR, %	Payment, £	APR, %	Payment, £	APR, %
0	250	0	125	0	83.3	0	62.5	0
2	255	3.2	130	3.6	88.3	3.7	67.5	3.7
4	260	6.5	135	7.2	93.3	7.3	72.5	7.4
6	265	9.8	140	10.8	98.3	11.0	77.5	11.0
8	270	13.2	145	14.4	103.3	14.6	82.5	14.6

(*Source:* Crabtree, 1984)

annual percentage rates. It is assumed that the person leasing the equipment derives no benefit from any sale proceeds and the one twelfth of the rentals are paid in advance (e.g. one advance payment for a three-year lease with quarterly rentals, or four advance payments for a four-year lease with monthly payments). A flat rate of 6 per cent for a three-year contract paid quarterly gives an annual percentage rate of 13.3 per cent (Fig. 9.9); paid half-yearly, the annual percentage rate increases further to 14.7 per cent because of the larger advance payment.

Even though the annual cost of borrowing is presented in a common format, the effect on hire purchase finance through varying the loan period and the effect of tax on finance leasing require separate consideration (see sections 9.5.4 and 9.5.5).

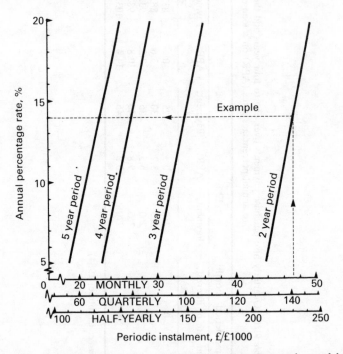

Fig. 9.9 A ready reckoner to transform monthly, quarterly, and half-yearly lease hire rentals over two-, three-, four- and five-year periods into the equivalent annual percentage rates, assuming one-twelfth of the rentals are fixed in advance (adapted from Bright, 1986).

9.5.2 Bank overdraft

Borrowing on a bank overdraft is generally the cheapest source of credit. The disadvantage is that funds lent on an overdraft are on ten days' call and can be withdrawn during a credit squeeze. While many bank managers will tolerate long periods of overdraft, they are under pressure to convert the debt to higher interest rate loans unless the account can be cleared of debt for any 30 days in the year. The interest charge on the overdraft is quoted as a percentage over the bank's Base Rate, usually $2\frac{1}{2}$–3 per cent, and is charged quarterly on the daily balance. As the overdraft rate fluctuates in the 12–16 per cent range, the borrower can become rather uncertain about the cost.

An overdraft is usually secured by a floating charge on all assets of the business and, should the borrower be unable to meet the conditions of the overdraft, there is a higher level of risk of a receiver being appointed, even though liquidation would leave the borrower with considerable equity.

9.5.3 Bank term loan

The term loan is a less risky method of financing a medium term investment because the loan is secured against some specific asset. In the worst eventuality, some financial restructuring will enable the business to survive.

The variable interest term loan is available for periods of two to seven years at an interest rate usually about 1 to 2 per cent above the overdraft rate. Some lenders charge an 'arrangement' fee which is typically 1–$1\frac{1}{2}$ per cent of the sum borrowed and this effectively increases the interest rate. On a five-year loan, for example, an 'arrangement' fee of 1 per cent is equivalent to increasing the annual percentage rate by about half a per cent. The loan capital is repayable at equally spaced intervals throughout the loan period or at the end, with interest being paid half-yearly. Unsecured term loans tend to cost considerably more and are, therefore, an even less desirable source of finance.

Some term loans allow the borrower to pay a floating rate of interest in the first instance, with an option to convert to a fixed rate for the remainder of the loan. This can be used to advantage during a period of temporarily high interest rates but it is always difficult to resist the temptation of staying with the lower variable rates and being caught out by the next financial crisis. It is also difficult to compare the cost of a fixed rate term loan with the cost of the variable rate applying to an overdraft. If a fall in the Base Rate is anticipated, then any fixed rate loan would look less attractive.

9.5.4 Hire purchase

One of the perceived advantages of hire purchase finance is that it is independent from the bank, so that the overdraft facility can be reserved for other capital needs where alternative finance is less readily available. This is particularly significant for a business with an overdraft near the agreed limit because it introduces the flexibility of using secondary finance to purchase machinery which could not otherwise be contemplated. An extra financial commitment of this nature does imply an increased risk for the whole business and should not be undertaken unless there is an adequate return in terms of extra income or reduced costs. In the event of the buyer failing to fulfil his contract, however, the hire purchase company has no call on the business as a whole. They can only repossess the goods. For this reason, an equity stake, by means of an initial deposit, is usual to dissuade the buyer from reneging on the contract.

Although the cost of hire purchase can be compared with a bank overdraft by means of their respective annual percentage rates, the extent to which the total finance cost is more or less expensive than overdraft also depends on the period of the loan. Differences between hire purchase costs and bank overdraft costs are given in Table 9.11. These differences apply to various hire purchase repayment frequencies for any given annual percentage rate, quarterly payments being taken as standard for bank overdrafts. For example, an interest-free hire purchase loan for one year is £60/£1000 cheaper than a bank overdraft at a nominal interest rate of 10 per cent, or £82/£1000 less than a bank overdraft at a nominal rate of 14 per cent. This means that one year's interest-free finance is equivalent to a price discount of 6–8 per cent in conjunction with a bank overdraft at a nominal interest rate of 10–14 per cent. In the same way, hire purchase finance at a flat rate of 4 per cent over four years is £54/£1000 cheaper

Table 9.11 Differences in finance costs between hire purchase and bank overdraft (quarterly payments). Negative figures indicate that hire purchase is cheaper than bank overdraft

Time period, month	Overdraft interest rate		Finance cost difference, £/£1000 at various flat rates of interest				
	Nominal, %	APR, %	0	2	4	6	8
12	10	10.4	− 60	− 41	− 22	− 3	16
	12	12.6	− 70	− 52	− 33	−15	3
	14	14.8	− 82	− 63	− 45	−27	−8
24	10	10.4	−104	− 68	− 32	3	40
	12	12.6	−123	− 87	− 52	−17	18
	14	14.8	−141	−106	− 72	−38	−3
48	10	10.4	−184	−119	− 54	12	77
	12	12.6	−215	−152	− 89	−27	36
	14	14.8	−244	−184	−123	−63	−2

(*Source*: Crabtree, 1984)

than a 10 per cent nominal rate overdraft, only marginally more expensive than one year's interest-free loan through hire purchase, and actually works out cheaper if the overdraft rates exceed 11 per cent. With a bank overdraft at a nominal rate of 14 per cent, hire purchase at a flat rate of 4 per cent over four years is equivalent to a price discount of 12.3 per cent (Table 9.11).

From the tabulated cost differences between hire purchase finance and bank overdraft, two main points are important. Firstly, low flat rates of interest over a short time period of one or two years appear more attractive than they really are, particularly if obtained at the expense of a price discount. Secondly, if the annual percentage rate for hire purchase is less than that for a bank overdraft, the total hire purchase cost is kept to a minimum by extending the repayment period for as long as possible.

9.5.5 Finance leasing

Finance leasing also provides independence from bank lending but, unlike hire purchase, the machine is never owned by the lessee, i.e. the farmer. Although this means that the farmer cannot claim capital allowances, he can use all the rental payments to reduce taxable income in the year in which they occur. These rental payments amount to much the same as the sum of the capital allowances and interest relief arising through the use of alternative sources of credit to finance the purchase of the machine, but the effect may differ because of variations in timing.

In the absence of taxable profits (whether through poor business performance, high interest charges or high levels of investment), leasing can be evaluated in much the same way as hire purchase. The annual percentage rates for the lease can be derived from Fig. 9.9 and compared with those for hire purchase in Table 9.10. For any given period, the lowest annual percentage rate indicates the cheapest source of finance. For example, a lease rental quotation of £96.00 per £1000, payable quarterly over three years, converts to an annual percentage rate of 11.2 per cent. In comparison with a bank overdraft at a nominal rate of 12 per cent, this lease is £26 per £1000 cheaper (Table 9.12).

Table 9.12 Differences in finance costs between leasing for an untaxed business and bank overdraft (quarterly lease payments over three years; first rented on signing). Negative figures indicate that leasing is cheaper than bank overdraft

Overdraft interest rate		Finance cost difference, £/£1000, for various lease rentals, £/£1000, and lease APR, %			
Nominal, %	APR, %	£94 9.4%	£96 11.2%	£98 13.0%	£100 14.8%
10	10.4	−24	− 7	28	49
12	12.6	−46	−26	− 5	25
14	14.8	−60	−40	−20	0

(*Source*: Crabtree, 1984)

9.6 ALTERNATIVES TO PURCHASE

The size, complexity and cost of certain farm machines has risen to the point where individual ownership may not always be fully justified. Instead of buying or leasing machinery, there are two further options to consider, namely:

○   **a contractor;**
○   **a machinery ring.**

A contractor is often relegated to provide only emergency cover in the event of mechanical failure of machinery at a critical time or to increase workrates after delays through bad weather. Whilst such occasions will always arise, planned use of a contractor or membership of a machinery ring may allow capital which would otherwise be tied up in machinery to be utilised more profitability elsewhere in the business.

9.6.1   Employing a contractor

The use of a contractor holds the prospect of easing the seasonal labour peaks, reducing the investment in machinery, and improving timeliness. Equally, however, there may be some concern that the loss of independence increases the business risk and places undue reliance on the verbal promise of the contractor to complete his obligations timeously and satisfactorily.

The greatest benefit from the use of a contractor is obtained by farmers who are working with a minimal labour force. The availability of an additional operator plus equipment during silage-making, for example, may be all that is required to avoid an extra man over the whole year. For the less mechanically minded, there is also the advantage of access to modern, efficient machinery, without the burden of high maintenance costs and without the management harassment of effecting rapid repairs during critical field operations.

A further benefit from the use of a contractor is the saving on heavy depreciation for machinery with a very short seasonal requirement. This saving releases capital for other investments in buildings, land improvement, machinery in regular use, or for purchasing livestock. One of the repercussions of the trend towards larger, more expensive self-propelled machines, such as combine harvesters, is that new machines are hard to justify below a certain scale of operation. The relative financial costs for three economy options – namely, employing a contractor, buying second hand equipment, or buying a new trailed machine – are shown in Fig. 9.10. A self-propelled combine harvester, capable of an overall rate of work of 1 ha/h and costing £29 000 new at 1982 prices, is shown to compete favourably with contract

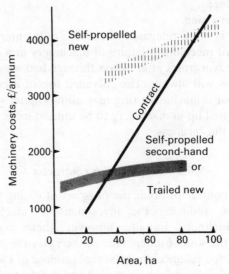

Fig. 9.10 The annual combining costs for various economy options compared with those for a new, self-propelled machine.

charges of £45/ha down to an annual area of about 80 ha. Good second hand machines (half price at three years old) are cheaper than contract charges down to annual areas of 30 ha but, below that crop area, the machines must be progressively older and kept longer with a greater risk of breakdowns. A new trailed combine harvester gives the reliability of a new machine with the costs of a used self-propelled unit, provided that a suitable tractor is available. Contracting is very competitive, especially where the available team is small and there is no one spare to drive an owned combine harvester.

The assurance of regular work on an annual basis enables the contractor to plan his commitments in advance. The availability of high capacity equipment also helps the contractor to achieve a level of operational timeliness as good as, if not better than, that with low capacity, less reliable owned equipment.

9.6.2 Machinery rings

On some farms there is likely to be a shortage of machinery; on others, there may be a surplus of machine capacity. The disparity is seldom eliminated by one or two co-operating farmers, whereas a larger group of participants can achieve a balance between supply and demand which is beneficial in different ways to all the individual members. The organisational structure which is required to formalise the arrangements between the participants is called a *machinery ring*.

- A **machinery ring** is an organisation, comprising a group of farmers and possibly contractors, which employs a manager to match the demands for machinery operations from some individual members with the supply of services available from others on a contractual basis.

The manager of the machinery ring schedules the use of machinery available in the ring, thereby ensuring both timeliness of operation and prompt payment. The members of the ring each pay a fee which relates to the size of the individual business, either on an area basis or on some other method which is common to all the participants. As the machinery ring does not support anything other than an office for the manager and his basic salary, the fees are only perhaps £20 to £30 per member. The manager also receives commission on every transaction, but all the machinery is owned by members of the ring and operated by the owner or by one of the owner's staff.

In most machinery rings, there are usually one or two contractors who only supply services; there may be a few farmers who only demand services and one or two who are inactive; the majority, however, both supply some services and demand others. The total machinery pool almost always has surplus capacity, so that work delayed by more than a day through machinery breakdowns is transferred to another member for completion. The formalising of neighbourly favours can also be used to keep the farm running during illness, accidents or even vacations. The income earned by those who supply machinery services is re-invested in new equipment and results in more frequent replacement of more heavily used machines, so that even the local machinery dealer is not adversely affected by the creation of a machinery ring.

9.7 FIXED COST CONTROL

High fixed costs in agriculture have been generated, to a large extent, by the farmers' traditional attitude towards investment in machines and buildings. Some of the complex reasons for this investment have little to do with the return on capital, for example:

minimising the payment of tax;

increasing the value of the holding;

making life easier;

following fashion;

ploughing back profits, without considering off-farm investments.

In many cases, the financial implications of the investments have not been fully evaluated.

Tax implications must always be carefully assessed and any advantages taken of possible allowances, but the prime consideration must be that the investment is beneficial to the business. Obsessive tax avoidance can come expensive. When stripped to the bare essentials, it very often involves spending £1 now in order to save 29 pence or 40 pence (depending on the marginal tax rate on profits) perhaps two years later when tax payments become due. If the investment expenditure of £1 is not essential, then it is far better to have the 60 pence or 71 pence to improve the cash balance and save on overdraft interest.

The *net worth* of the business as a whole is the real indicator of financial viability and is the figure which is scrutinised during any review of overdraft facilities.

● The **net worth** of a business is its market value. Investments which improve the quality of life, which enhance the stature of the entrepreneur within the farming community, and which are reserved exclusively for the farm business are privileges to be treated like luxury goods – highly desirable for those who can afford them.

9.8 SUMMARY

Fixed costs are business overheads which do not alter with small changes in output.

The gang workday chart is used to plan the allocation of labour resources at peak periods.

Arable farm planning for maximum profitability involves linear programming techniques to determine the optimum cropping plan, team size and machinery complement.

The optimum period of machine ownership is primarily

dependent on the rate of depreciation and the annual repair costs.

The whole stock method of scheduled machinery replacement ensures a uniform annual level of re-investment in new machinery, and replacement of all machines in rotation over a reasonable timespan.

The standard method of comparing interest rates is by the annual percentage rate.

The main sources of finance for machinery purchase are bank overdraft, bank loans, hire purchase, and leasing.

Employing a contractor or being a member of a machinery ring can release capital which would otherwise be tied up in machinery for use elsewhere in the business.

The future profitably of arable farms is largely dependent on reducing the level of fixed costs.

APPENDIX A1
COMPARISON BETWEEN
CONVENTIONAL AND
REVERSIBLE PLOUGHING

A1.1 INTRODUCTION

Identifying the optimum fieldwork pattern combines an exercise in method study to expedite marking out the field (section 2.3.5 and Fig. 2.10) with an analysis of the factors influencing machine rates of work (section 3.8.3 and Fig. 3.11).

A1.2 FIELD EFFICIENCY FOR CONVENTIONAL PLOUGHING

A1.2.1 Setting out headlands

Travel once round the field using the tractor for transport and stopping at intervals to pace the headland width, mark with a spade and walk back to the tractor.

Travel once round the field at ploughing speed to score the headlands and sidelands.

No. of markers
$$= 2\frac{l_f + W_a}{x} + 2$$

Walking time, t_{w1}
$$= \frac{4 W_h}{1000 V_w}\left[\frac{l_f + W_a}{x} + 2\right] \quad \ldots [A1.1]$$

Tractor time travelling, $t_{i_1} = \frac{2(l_f + W_a)}{1000 V_i} \quad \ldots [A1.2]$

Tractor time ploughing, $t_{p_1} = \frac{2(l_f + W_a)}{1000 V_p} \quad \ldots [A1.3]$

A1.2.2 Setting out opening ridges

Travel down one sideland using the tractor for transport and stopping at intervals to pace across to the first opening ridge, mark with a ranging pole and walk back to the tractor. Travel from sideland along headland to first opening ridge.

Travel along first opening ridge at plough speed to score it and stopping at intervals to pace across to next opening ridge, mark with a ranging pole and walk back to the tractor. Travel across headland to next opening ridge and repeat.

Width of gathered half land $\quad = \dfrac{W_a}{(2\,N_r - 1)}$

No. of markers/opening ridge $\quad = 4$

Walking distance from sideland to 1st opening ridge, return

$$= \frac{W_a}{(2\,N_r - 1)}$$

Walking distance between subsequent opening ridges, return

$$= \frac{4\,W_a}{(2\,N_r - 1)}$$

$\therefore$ Total walking distance for N_r opening ridges with 4 markers/furrow, return

$$= 4\,\frac{W_a}{(2\,N_r - 1)} + \frac{4(N_r - 1)\,W_a}{(2\,N_r - 1)}$$

$$= \frac{4(4\,N_r - 3)W_a}{(2\,N_r - 1)}$$

$\therefore$ Walking time, t_{w2}
$$= \frac{4(4\,N_r - 3)\,W_a}{1000(2\,N_r - 1)\,V_w}$$

$$\cong \frac{8\,W_a}{1000\,V_w} \qquad \ldots \text{[A1.4]}$$

Tractor time travelling and ploughing, $t_{i2} + t_{p2}$

$$= \frac{l_f}{1000\,V_i} + \frac{W_a}{2 \times 1000(2\,N_r - 1)V_i}$$

$$+ \frac{4(N_r - 1)W_a}{2 \times 1000(2\,N_r - 1)V_i} + \frac{N_r\,l_f}{1000\,V_p}$$

$$= \frac{(4 N_r - 3)W_a}{2 \times 1000(2 N_r - 1)V_i} + \frac{l_f}{1000 \, V_i} + \frac{N_r \, l_f}{1000 \, V_p}$$

$$\cong \frac{W_a}{1000 \, V_i} + \frac{(N_r + 1)l_f}{1000 \, V_p}$$

where: $t_{i2} \cong \dfrac{W_a}{1000 \, V_i}$ $\qquad \ldots$ [A1.5]

and: $t_{p2} \cong \dfrac{(N_r + 1)l_f}{1000 \, V_p}$ $\qquad \ldots$ [A1.6]

A1.2.3 Ploughing

Plough round each opening ridge until half land complete, travel to next opening ridge until all gathered half lands complete, then travel back across the field completing each cast half land in turn.

Tractor time ploughing one pass $\quad = \dfrac{l_f}{1000 \, V_p}$

Number of passes $\quad = \dfrac{W_a}{W_p}$

$\therefore$ Total time ploughing, $t_{p3} \quad = \dfrac{l_f \, W_a}{1000 \, W_p \, V_p}$

$$\ldots \text{[A1.7]}$$

Extra tractor time ploughing opening ridges, $t_{p4} = \dfrac{2 \, N_r \, l_f}{1000 \, V_p}$

$$\ldots \text{[A1.8]}$$

Extra tractor time ploughing finishes, $t_{p5} = \dfrac{(N_r - 1)l_f}{1000 \, V_p}$

$$\ldots \text{[A1.9]}$$

Average tractor turning time on headland

$$= \frac{W_a}{2 \times 1000(2 \, N_r - 1)V_i}$$

$\therefore$ Total tractor turning time, t_{i3}

$$= \frac{(W_a)^2}{2 \times 1000(2\ N_r - 1)W_p\ V_i} \qquad .\,.\ \text{[A1.10]}$$

Extra tractor travelling time between half lands to complete field, t_{i4}

$$= \frac{3(N_r - 1)W_a}{1000(2\ N_r - 1)V_i} + \frac{W_a}{1000(2\ N_r - 1)V_i}$$

$$= \frac{(3\ N_r - 2)W_a}{1000(2\ N_r - 1)V_i}$$

$$\cong \frac{W_a}{1000\ V_i} \qquad \ldots \text{[A1.11]}$$

A1.2.4 Ploughing headlands and sidelands

Tractor time ploughing, $t_{p6} = \dfrac{2\ W_h(l_f + W_a)}{1000\ W_p\ V_p}$

$$\ldots \text{[A1.12]}$$

Tractor time turning corners of field on headlands, t_{t1}

$$= \frac{4\ W_h\ t_m}{W_p} \qquad \ldots \text{[A1.13]}$$

A1.2.5 Optimum number of opening ridges

The optimum number of opening ridges is achieved when the field efficiency is a maximum. The field efficiency is the ratio of the time spent ploughing to the total time in the field. The time spent setting out the field, t_{w2} and t_{i2}; the time travelling between lands, t_{i4}; the time completing the headland and sidelands, t_{w1}, t_{i1}, t_{p1}, t_{p6} and t_{t1} are not considered in the determination of the number of opening ridges because this time is constant regardless of the number of opening ridges.

Thus, pattern time ratio is:

$$R_p = \frac{t_{p3}}{t_{p3} + t_{i3} + t_{p2} + t_{p4} + t_{p5}}$$

$$= \frac{\dfrac{l_f\,W_a}{1000\,W_p\,V_p}}{\dfrac{l_f\,W_a}{1000\,W_p\,V_p} + \dfrac{(W_a)^2}{2\times 1000(2\,N_r - 1)W_p\,V_i} + \dfrac{4\,N_r\,l_f}{1000\,V_p}}$$

$$= \frac{(4\,N_r - 2)l_f\,W_a\,V_i}{(4\,N_r - 2)\,l_f\,W_a\,V_i + (W_a)^2\,V_p + 4(4\,N_r - 2)N_r\,l_f\,W_p\,V_i}$$

Let $R_p = \dfrac{u}{v}$; then $\dfrac{d(R_p)}{d(N_r)} = \dfrac{u\,\dfrac{dv}{d(N_r)} - v\,\dfrac{du}{d(N_r)}}{u\,v}$

When $\dfrac{d(R_p)}{d(N_r)} = 0$, R_p is a maximum

$\therefore\ u\,\dfrac{dv}{d(N_r)} = v\,\dfrac{du}{d(N_r)}$

$\dfrac{du}{d(N_r)} = 4\,l_f\,W_a\,V_i$

$\dfrac{dv}{d(N_r)} = 4\,l_f\,W_a\,V_i + (32\,N_r - 8)\,l_f\,W_p\,V_i$

$\therefore\ (4\,N_r - 2)l_f\,W_a\,V_i\,[4\,l_f\,W_a\,V_i + (32\,N_r - 8)l_f\,W_p\,V_i]$
$= 4\,l_f\,W_a\,V_i\,[(4\,N_r - 2)l_f\,W_a\,V_i + (W_a)^2\,V_p]$
$\qquad + 4(4\,N_r - 2)N_r\,l_f\,W_p\,V_i$

$\therefore\ (4\,N_r - 2)(32\,N_r - 8 - 16\,N_r)l_f\,W_p\,V_i = 4(W_a)^2 V_p$

$\therefore\ (4\,N_r - 2)^2 l_f\,W_p\,V_i = (W_a)^2 V_p$

$\therefore\ 4\,N_r - 2 = \left[\dfrac{(W_a)^2 V_p}{l_f\,W_p\,V_i}\right]^{0.5}$

$\therefore N_r = 0.5 + \left[\dfrac{(W_a)^2 V_p}{16\,l_f\,W_p\,V_i}\right]^{0.5}$ $\qquad\qquad$ . . . [A1.14]

A1.2.6 Field efficiency

$FE_c = (t_{p3} + t_{p6})/[(t_{p3} + t_{p6}) + (t_{p1} + t_{p2} + t_{p4} + t_{p5}) + t_{i3}$
$\qquad + (t_{i1} + t_{i2} + t_{i4}) + (t_{w1} + t_{w2}) + t_{t1}]$

$\qquad\qquad$ . . . [A1.15]

A1.3 FIELD EFFICIENCY FOR REVERSIBLE PLOUGHING

A1.3.1 Setting out headlands

Travel across headland using tractor for transport and stopping at intervals to pace the headland width, mark with a spade and walk back to the tractor.

Travel back across headland at ploughing speed to score the headland.

Travel at transport speed length of field to opposite headland, repeat marking procedure then travel back to starting point.

No. of markers $= 2(W_a/x + 1)$

Walking time, t_{w1} $= \dfrac{4\ W_h(W_a/x + 1)}{1000\ V_w}$ ... [A1.16]

Tractor travelling time, t_{i1} $= \dfrac{2(l_f + W_a)}{1000\ V_i}$ ... [A1.17]

Tractor ploughing time, t_{p1} $= \dfrac{2\ W_a}{1000\ V_p}$ ... [A1.18]

A1.3.2 Ploughing

Plough full width of field including sidelands.

Tractor ploughing time, $t_{p2} = \dfrac{l_f\ (W_a + 2\ W_h)}{1000\ W_p\ V_p}$ ... [A1.19]

Tractor turning time, t_{t1} $= \dfrac{(W_a + 2\ W_h)\ t_m}{W_p}$ ... [A1.20]

A1.3.3 Ploughing headlands

Plough headlands, turning at each end and travelling between headlands at transport speed.

Tractor ploughing time, $t_{p3} = \dfrac{2(W_h\ W_a)}{1000\ W_p\ V_p}$ ... [A1.21]

Tractor travelling time, $t_{i2} = \dfrac{2\, l_f}{1000\, V_i}$ ... [A1.22]

Tractor turning time, $t_{t2} = \dfrac{2\, W_h\, t_m}{W_p}$ ... [A1.23]

A1.3.4 Field efficiency

$$FE_r = \frac{(t_{p2} + t_{p3})}{(t_{p2} + t_{p3}) + t_{p1} + (t_{i1} + t_{i2}) + t_{w1} + (t_{t1} + t_{t2})}$$

... [A1.24]

LIST OF REFERENCES

ADAS 1970 The utilisation and performance of combine harvesters, 1969. Farm Mech. Studies No. 18, Agric. Dev. Adv. Serv. Min. Agric. Fish. Food.

ADAS 1972 The utilisation and performance of potato harvesters, 1971. Farm Mech. Studies No. 24, Agric. Dev. Adv. Serv. & Potato Marketing Board. Min. Agric. Fish. Food.

ADAS 1972 A joint ADAS/NIAE study of utilisation, performance and tyre and track costs 1969–70. Agric. Dev. Adv. Serv. Min. Agric. Fish. Food.

ADAS 1974 The utilisation and performance of sugar beet harvesters, 1973. Farm Mech. Studies No. 26. Agric. Dev. Adv. Serv. Min. Agric. Fish. Food.

ADAS 1976 The effect of different types of potato planter on crop yield and profitability. Short-Term Leaflet No. 182, Agric. Dev. Adv. Serv. Min. Agric. Fish. Food.

ADAS 1983 Forage harvesting systems survey, 1983. Agric. Dev. Adv. Serv. Min. Agric. Fish. Food.

Anderson J L 1986 Profitability of farming in South-East Scotland 1984/85. Economics and Management Series No. 20, East Scotl. Coll. Agric., Edinburgh.

Anon 1877 *Notes and sketches illustrative of northern rural life in the eighteenth century*. David Douglas, Edinburgh.

Anon 1968 *A century of agricultural statistics Great Britain 1866–1966*. HMSO, London.

Anon 1977 Steering made clear. *Power Farming*, Oct., pp 10–11, 13, 15.

Anon 1981 *AEA data book*. Agric. Engrs Assn Ltd, London.

Anon 1982 Measure the risks before you spray. *Farmer's Weekly* Extra Tank Mixes, Feb. 26, pp 8–9.

Anon 1983 Scope for reduction of inputs in farming. Report of a Joint Working Group representing Agric. Res. Council, Min. Agric. Fish. Food, Dept Agric. Fish. Scotl., and Dept Agric. N. Ireland.

Anon 1984 Ten tractors on test: the view from the driving seat. *Power Farming*, Oct., pp 22–3, 25, 27.

Anon 1985 *Output and utilisation of farm produce in the United Kingdom 1978–84*. HMSO, London.

Anon 1986 *Agricultural statistics United Kingdom 1984*. HMSO, London.

Anon 1986 *Yearbook of agricultural statistics 1985*. Statistical Office of the European Communities, Luxembourg.

Anon 1986 Reliability: the real-life evidence. *Sunday Times*, 30 Mar., p 76.

Arthey D 1986 Quality of agricultural products. *Agric. Engr* **41**: 91–6.

ASAE 1982 *Agricultural engineering year book*. Am. Soc. agric. Engrs, St Joseph, Mich.

ASHRAE 1985 *Handbook of fundamentals*. Am. Soc. Heating, Refrigerating and Air-conditioning Engrs, p 8.17.

Audsley E 1981 An arable farm model to evaluate the commercial viability of new machines or techniques. *J. agric. Engng Res.* **26**: 135–49.

Audsley E. Boyce D S 1974 A method of minimising cost of harvesting and high temperature grain drying. *J. agric. Engng Res.* **19**: 173–89.

Audsley E, Wheeler J 1978 The annual cost of machinery calculated using actual cash flows. *J. agric. Engng Res.* **23**: 189–201.

Ayres R M, Waizeneker J 1978 A practical approach to vehicle replacement. *Chartered Mech. Engr*, Sept., pp 73–5.

Baillie W F, Brown W T 1971 Summary report of tests on tractors in the 60–70 hp range. Australian Tractor Testing Committee, Dept agric. Engng, Univ. of Melbourne.

Barrett F M 1982 Arable work scheduling and projection: an adviser's method. *Agric. Engr* **37**: 120–2.

Barrett F M 1985 Machinery replacement. ADAS Bull. for Oxford Div., Agric. Dev. Adv. Serv.

Barrington J 1984 *Red sky at night*. Michael Joseph, London.

Barton G A 1972 Marketing farm durables. In Bateman D I (ed.) *Marketing management in agriculture*. University College of Wales, Aberystwyth.

Bell J W 1977 The cost of combine breakdowns and delayed harvesting. Farm Management Review, No. 10. North Scotl. Coll. Agric., Aberdeen.

Bender A E 1986 A scientific basis for a healthy diet. Royal agric. Soc. England & Agric. Dev. Adv. Serv. Conf. entitled: Dietary health and implications for agriculture, Natn. agric. Centre, Kenilworth.

Bird L G 1978 Farm odours and the weather. ADAS Meteorologist, Min. Agric. Fish. Food, Harrogate.

Blackmore R A, Tyldesley J B 1974 Sequences of dry days for grass conservation in Cumbria, Northumberland and Durham. Agric. Memo. No. 674, Meteorological Office.

Bottoms D J 1976 An introduction to some human performance measures with particular reference to tractor driving. File Note No. 006, Natn. Inst. agric. Engng. Silsoe.

Bowers W 1975 *Machinery management*. Fundamentals of machine operation. John Deere Service Publications, Moline, Illinois.

Bowers W, Hunt D R 1970 Applications of mathematical formulas to repair cost data. *Trans Am. Soc. agric Engrs* **13**: 806–9.

Bowman J C 1986 Environmental and governmental pressure on agriculture. *Agric. Engr* **41**: 77–83.

Bright G 1986 The current economics of machinery leasing. *Farm Management* **6**: 105–12.

Brixius W W, Wismer R D 1978 The role of slip in traction. Paper No. 78–1538, Am. Soc. agric. Engrs, St Joseph, Mich.

Brown F R, Charlick R M 1972 An interpretation of the effect of conditioning in relation to meteorological data for two regions. Subject Day Paper No. 4, Natn. Inst. agric. Engng, Silsoe.

Buckett M 1981 *An introduction to farm organisation and management*. Pergamon Press, Oxford.

Burstall A F 1963 *A history of mechanical engineering*. Faber and Faber, London

Butterworth B, Nix J 1983 *Farm mechanisation for profit*. Granada Publishing, London.

Camm B M 1985 Choice of borrowing. *Farm Management* **5**:433–42.

Campbell R J 1982/83 Planning to employ a contractor. *Farm Management* **4**: 471–8.

Carter E S 1984 Restrictions on the farm business. *Farm Management* **5**: 237–45.

Cermak J P, Ross P A 1977 Labour requirements of beef housing: check list. *Farm Building Progress*, No. 47, 9–10.

Cermak J P, Ross P A 1978 Airborne dust concentrations associated with animal housing tasks. *Farm Building Progress*, No. 51, 11–5.

Cermak J P, Ross P A 1979 Noise deafness in livestock housing. *Farm Building Progress*, No. 58, 5–7.

Cermak J P, Ross P A 1980 Ergonomic data for some tasks in cattle and pig housing. *Farm Building Progress*, No. 61, 7–11.

Cermak J P, Ray R D, Clark J J 1981 Spatial requirements of some animal housing tasks. *Agric. Manpower*, No. 3, 21–3.

Chancellor W J, Thai N C 1984 Controlling tractor engine and transmission automatically. *Agric. Engng* **65**(5): 24–9.

COMA 1983 Diet and cardiovascular disease. Report on Health and

Social Subjects No. 28. Committee on Medical Aspects of Food Policy. HMSO.

Crabtree J R 1984 Machinery finance options. *Farm Management* **5**: 285–93.

Christiansen E M 1953 Physiological evaluation of work in the Nykroppa iron works. In Floyd W F, Welford A T (eds) *Fatigue*, p 93.

Crowe S 1971 Agriculture and landscape, *Outlook on Agriculture* **6**: 291–6.

Culpin C 1975 *Profitable farm mechanisation*. Crosby Lockwood Staples, 3rd Edition, London.

Currie R M 1972 *Work study*. Revised by Faraday J E. Pitman Publishing/Management Publications, London.

Davis W M 1965 How good are farm tractor transmissions? *Agric. Engng* **46**: 380–2, 390.

Dupuis H 1959 Effect of tractor operation on human stresses. *Agric. Engng* **40**: 510–25.

Dwyer M J, Comely D R, Evernden D W 1975 Development of the N.I.A.E. handbook of agricultural tyre performance. *Proc. 5th Int. Conf., Int. Soc. Terrain-Vehicle Systems*, *Detroit*, pp 679–99.

Dwyer M J, Evernden D W, McAllister M 1976 Handbook of agricultural tyre perfomance. Report No. 18 Natn. Inst. agric. Engng, Silsoe.

Dwyer M J 1978 Maximising agricultural tractor performance by matching weight, tyre size and speed to the power available. *Proc. 6th Int. Conf., Int. Soc. Terrain-Vehicle Systems*, *Vienna*, pp 479–93.

Dwyer M J 1982 Tyres: the most important factor. *Power Farming*, July, pp 16–18, 23.

Dwyer M J 1984 Computer models to predict the performance of agricultural tractors on heavy draught operations. *Proc. 8th Int. Conf., Int. Soc. Terrain-Vehicle Systems*, *Cambridge*, pp 933–52.

Dyer J A, Baier W 1979 An index for soil moisture drying patterns. *Can. agric. Engng* **21**: 117–18.

Easterby R 1975 Psychological aspects of agricultural ergonomics. *Agric. Engr* **30**: 112–14.

Edlin H L 1960 Silviculture in the New Forest. In *The New Forest*. Galley Press, London.

Edwards G J 1968 The requirements of tractor transmissions. *J. Proc. I. Agr. E.* **23**: 202–11.

Elbanna E B 1986 Agricultural machinery selection: soil strength and operational timeliness. PhD thesis (unpubl.), Edinburgh Univ.

Elbanna E B, Witney B D 1987 Cone penetration resistance equation as a function of the clay ratio, soil moisture content and specific weight. *J. Terramechanics* **24**: 41–56.

van Elderen E 1980 Models and techniques for scheduling farm operations: a comparison. *Agric. Systems* **5**: 1–17.

Elrick J D, Bright G A 1981 A simple approach to inflation-proof budgets. Tech. Note No. 285EM, East Scotl. Coll. Agric., Edinburgh.

Elrick J D 1982 How to choose and use combines. Publ. No. 88, Scott. agric. Coll.

Eradat Oskoui K 1986 Days available for fieldwork. Tech. Note. No. 97, Scott. agric. Coll.

Eradat Oskoui K, Witney B D 1982 The determination of plough draught. I – Prediction from soil and meteorological data with cone index as the soil strength parameter. *J. Terramechanics* **19**: 97–106.

Eradat Oskoui K, Rackham D H, Witney B D 1982 The determination of plough draught. II – The measurement and prediction of plough draught for two mouldboard shapes in three soil series. *J. Terramechanics* **19**: 153–64.

Eriksson H 1973 Climatic requirements in tractor cabs. Rep. No. 9, Jordbrukstekniska Institutet, Uppsala.

Eurostat 1979 The agricultural situation in the Community. Community survey on the structure of agricultural holdings, 1975. Vol II.

Fairless M *The Roadmender*. Collins.

Foxall G R 1979 Farmers' tractor purchase decisions: a study of interpersonal communication in industrial buying behaviour. *Eur. J. Marketing* **13**: 299–308.

Friedlander J N 1979 Vehicle replacement. *Chartered Mech. Engr*, Jan., p 51.

Funk T F 1972 Farmer buying behaviour: an integrated review of literature. Working paper No. AE/72/16, School of agric. Econ. and Extension Education, Ontario Agric. Coll., University of Guelph.

Gee-Clough D 1980 Selection of tyre sizes for agricultural vehicles. *J. agric. Engng Res.* **25**: 261–78.

Gill A H 1971 Variation in the repair costs of tractors, combine harvesters and balers. Misc. Study No. 50, Dept agric. Econ. & Man., Reading Univ.

Gladwell G 1985 Labour and machinery analysis and planning. *Farm Management* **5**: 365–72.

Glasbey C A, McGechan M B 1986 The assessment of combining workdays criteria and forecasting models. *J. agric. Engng Res.* **33**: 23–31.

Gloyne R W, Armour D G Sequences of dry days in S. W. Scotland within the three-month period May to July (inclusive). Agric. Memo. No. 530, Meteorological Office.

Green J O, Corrall A J, Terry R A 1971 Grass species and varieties –

relationship between stage of growth, yield and forage quality. Tech. Report No. 8, Grassland Research Institute.

Hagger S 1986 The ownership costs of tractors. BSc (Hons) dissertation (unpubl.), Univ. of Edinburgh.

Handley J E *Scottish farming in the eighteenth century*, Faber & Faber Ltd, London.

Harley I D 1982 Factors influencing machinery selection for cereal crop establishment in the spring. BSc (Hons) dissertation (unpubl.). Edinburgh Univ.

Hartman M A, Baird R W, Pope J B, Knisel W G 1960 Determining rainfall-runoff-retention relationships. Bull. No. MP-404, Texas agric. Exp. Stn.

Health Education Council 1983 Proposals for nutritional guidelines for health education in Britain. Discussion paper, Natn. Adv. Comm. Nut. Education.

Hick W E 1952 On the rate of gain of information. *Quart. J. exp. Psychol.* **4**: 11–26.

Hockey W S 1961–62 Tractor mounted implements and adaptations. Symposium on Agricultural Tractors. *Proc. Automobile Div., I. Mech. E.* (4): 158–66.

Hole C 1978 *A dictionary of British folk customs*. Paladin Books, Granada Publishing, London.

Howe S, Bird M 1985 Nine leading tractors . . . what their owners think. *Power Farming* Update, July, pp U6–U9.

HSE 1980 Prevention of tractors overturning. Agric. Safety Leaflet No. AS22, Health and Safety Executive.

HSE 1980 Noise. Agric. Safety Leaflet No. AS8, Health and Safety Executive.

Humphreys G Merry ploughman of old England. The Massey Ferguson Newspaper for Farmers. Modern Farmer.

Humphreys G 1984 Homage to the Corn-God. The Massey Ferguson Newspaper for Farmers. Modern Farmer **22**(8).

Hunt D R, Fujii K 1976 Repair and maintenance costs by machinery categories. Paper No. 76–1507, Am. Soc. agric. Engrs, St Joseph, Mich.

Hunt D R 1977 *Farm power and machinery management*. Iowa State Univ. Press.

Inns F M, Kilgour J 1978 *Agricultural tyres*. Dunlop Ltd, King St, London.

Inns F M 1985 Some design and operational aspects of 3-link implement attachment systems. *Agric. Engr* **40**: 136–44.

ISO 1974 Guide for the evaluation of human exposure to whole body vibration. ISO 2631. International Standards Organisation.

Jeffrey W A 1981 Power ratings of diesel engined farm machinery. Tech. Note No. 268 E/C/A, East Scotl. Coll. Agric., Edinburgh.

Jenkins J E E, Storey I F 1975 Influence of spray timing for the control of powdery mildew on the yield of spring barley. *Pl. Path.* **24**: 125–34.

Jewett D W 1983 The mechanisation muddle – which way forward? *Agric. Engr* **38**: 87–90.

Kachatryan Kh A 1970 External information and the work of operators of farm machinery. *Doklady Vaskhnil* **6**: 40–2. Trans. No. 258. Natn. Inst. agric. Engng, Silsoe.

Kepkay L L 1972 Turning techniques. *Power Farming*, Feb., pp 54–5.

Kirkpatrick D 1979 Machinery costs relating to replacement policy. BSc (Hons) dissertation (unpubl.), Edinburgh Univ.

Knisel W G, Baird R W 1969 Runoff volume prediction from daily climatic data. *Water Resources Res.* **5**: 84–94.

Kofoed S S 1984 An automatic coupling device for tractor pto drivelines. *J. agric. Engng Res.* **30**: 347–52.

Lewis M 1981 Intake prediction equations for forage based diets. Agric. Dev. Adv. Serv. Nutrition Chemists Tech. Conf., Min. Agric. Fish. Food.

Liljedahl J B, Carleton W M, Turnquist P K, Smith D W 1979 *Tractors and their power units* (3rd edn). J Wiley, New York.

Link D A 1967 Activity network techniques applied to a farm machinery selection problem. *Trans Am. Soc. agric. Engrs* **10**: 310–17.

List R J 1985 *Smithsonian meteorological tables* (6th rev). Smithsonian Inst.

Lucas N 1983 Four-wheel drive. *Power Farming*, July, pp 31–3, 35.

McGechan M B 1984 Work study on combine harvesters using automatic telemetry equipment. *Agric. Engr* **39**: 131–8.

McGechan M B 1985 A parametric study of cereal harvesting models. I: Critical assessment of measured data on parameter variability. *J. agric. Engng Res.* **31**: 149–58

McGechan M B 1985 Initial experience with using the NIAE forage conservation system simulation model to compare Scottish and English sites. Dept. Note No. Sin/434, Scott. Inst. Agric. Engng, Penicuik.

Mackay A J M 1983 Machinery rings. Tech. Note No. 325M, East Scotl. Coll. Agric., Edinburgh.

McLean K A 1980 *Drying and storing combinable crops*. Farming Press Ltd, Ipswich.

McLeod D J, Staffurth C 1968 Resource allocation. In Thornley G (ed.) *Critical path analysis in practice.* Tavistock Publications, London.

McRae D C 1979 Mechanical handling damage. Project 55 – potatoes. Central Council for Agricultural and Horticultural Co-operation, London. '

MAFF *Business statistics.* Min. Agric. Fish. Food.

Malcolm D G, DeGarmo E P 1953 Visual inspection of products for surface characteristics in grading operations. USDA Marketing Rep. No. 45.

Matthews J 1967 Some measurements of carbon monoxide pollution of glasshouse atmosphere during mechanised cultivation. Note No. 7, Natn. Inst. agric. Engng, Silsoe.

Mertins K-H 1978 Scientific approach to big tractor steering systems. *Power Farming*, Feb., p 52.

Monk A S, Morgan D D V, Morris J, Radley R W 1984 The cost of farm accidents. Occas. Paper No. 13, Silsoe Coll.

Morling R W 1979 Agricultural tractor hitches – analysis of design requirements. Distinguished Lecture Series, Am. Soc. agric. Engrs, St Joseph, Mich.

Nation H J 1978 Logistics of spraying with reduced volumes of spray and higher vehicle speeds. *BSRAE Association Leaflet – Weeds.* British Crop Protection Council, pp 641–8.

NIAE 1983 Report on test of John Deere 4450 tractor with mechanical front wheel drive. Rep. No. 694, Natn. Inst. agric. Engng, Silsoe.

NIAE 1985 Report on test of Ford 8210 tractor. Rep. No. 705, Natn. Inst. agric. Engng, Silsoe.

Nix J 1986 *Farm management pocketbook* (17th edn). Wye Coll. (London Univ.).

Norvell D W 1980 Farmers' buying behaviour as related to the purchasing of farm implements in Kansas. Monograph 2, Agric. Exp. Stn, Kansas State Univ., Manhattan.

Owen G M, Hunter A G M 1983 A survey of tractor overturning accidents in the United Kingdom. *J. occup. Accidents* **5**: 185–93.

Palmer J 1984 Collecting data from tractors in everyday use. *Agric. Engr* **39**: 38.

Patterson D E, Chamen W C T, Richardson C D 1975 Perennial experiments with tillage systems to improve the economy of cultivations for cereals: 4th year – 1974/75 experiments. Dept Note No. DN/Cu/615/1260, Natn. Inst. agric. Engng, Silsoe.

Philips P R, O'Callaghan J R 1974 Cereal harvesting – a mathematical model. *J. agric. Engng Res.* **19**: 415–33.

Pierce L T 1960 A practical method of determining evapotranspiration from temperature and rainfall. *Trans Am. Soc. agric. Engrs* **3**: 77–81.

Pierce L T 1966 A method for estimating soil moisture under corn, meadow, and wheat. Res. Bull. No. 988, Ohio Agric. Res. Dev. Centre, Wooster, Ohio.

Pigott B 1960 The New Forest Commoners. In *The New Forest*. Galley Press, London.

Purcell W F M 1980 The human factor in farm and industrial design. Distinguished Lecture Series. Am. Soc. agric. Engrs, St Joseph, Mich.

RASE 1984 *Agricultural tractors 1985*. Royal agric. Soc. England.

Reynolds J 1970 *Windmills and watermills*. Hugh Evelyn, London.

Ricketts C J, Weber J A 1961 Tractor engine loading. *Agric. Engng* **92**: 236–9, 250.

Rosseger R, Rosseger S 1960 Health effects of tractor driving. *J. agric. Engng Res.* **5**: 241–74.

Rutherford I 1976 Machinery costings in times of inflation – a standard method. Occas. Notes on Mechanisation No. 246, Agric. Dev. Adv. Serv. Min. Agric. Fish. Food.

Rutherford I 1977 The faults – the utilisation and performance of field crop sprayers. *Agric. Engr* **32**: 40–4.

SAC 1979 Farm dusts and your health. Publ. No. 52, Scott. agric. Coll.

SAC 1986 *Farm management handbook* (7th edn). Publ. No. 168, Scott. agric. Coll.

Schrader W 1982 Dial and switch implement hitch. *Power Farming*, Sept, pp 39, 41.

Scott A N, Audsley E 1981 Relating tractor specifications to ploughing performance. Div. Note No. DN/1056 (unpubl.). Natn. Inst. agric. Engng, Silsoe.

Shepherd W 1958 Moisture relations of hay species. *Aus. J agric. Res.* **9**: 436–45.

Shoup W D 1983 Conjoint analysis for forecasting tractor design features. *Trans Am. Soc. agric. Engrs* **26**: 1282–6.

Singleton W T 1974 *Man machine systems*. Penguin Education Series.

Sjøflot L 1982 Safe and easy coupling of tractor implements. *Proc. 8th Congress Int. Ergonomics Assoc.*, pp 502–3.

Smith K, Nielsen V 1983 Agricultural odours: some control procedures. *Agric. Engr* **38**: 21–4.

Smith E A, Bailey P H, Ingram G W 1981 Prediction of the field moisture content of barley and wheat by commonly used drying equations. *J. agric. Engng Res.* **26**: 171–8.

Soane B D, Blackwell P S, Dickson J W, Painter D J 1980–81 Compaction by agricultural vehicles: a review. II: Compaction under tyres and other running gear. *Soil & Tillage Res.* **1**: 373–400.

Söhne W 1959 Untersuchungen über die Form von Pflugkörpern bei ehröhter Fahrgeschwindigkeit. (Investigations on the shape of plough bodies for high speeds.) Trans. No. 87. Natn. Inst. agric. Engng, Silsoe. *Grundlagen der Landtechnik* **11**: 22–39.

Spackman E A 1983 Spray-occasions: June 1982 to May 1983. British Crop Protection Council, 20th Review of Herbicide Usage, pp 4–10.

Spackman E A, Barrie I A 1981 Spraying occasions determined from meteorological data during the 1980–81 season at 15 stations in the UK and comparison with 1971–80. British Crop Protection Council, 18th Review of Herbicide Usage, pp 3–12.

Spatz G, van Elmern J, Lawrynowicz R 1970 Der trocknugsverlanf von heu im Frieland. (Variation of drying rate of hay in the field.) *Bayerisches Landwirtschaftliches Jahrbuch* **47**: 446–64. (English Trans. No. 315. Natn. Inst. agric. Engng, Silsoe.)

Spencer H B, Bowden P J, Smith E A, Graham R, Glasbey C A 1985 The effect of conditioning on silage wilting – a field trial (1984). Dept. Note. No. SIN/435, Scott. Inst. agric. Engng, Penicuik.

Spencer H B, Bowden P J, Smith E A, Graham R, Glasbey C A 1985 The effect of conditioning on hay drying – a field trial (1984). Dept. Note. No. Sin/436, Scott. Inst. agric. Engng, Penicuik.

Spencer T D 1961 Effects of carbon monoxide on man and canaries. *Ann. occup. Hyg.* **5**: 231–40.

Stayner R M, Bean A G M 1975 Tractor ride investigations: a survey of vibrations experienced by drivers in field work. Dept Note No. DN/E/578/1445 (unpubl.), Natn. Inst. agric. Engng, Silsoe.

Stansfield J R 1974 Fuel in British Agriculture. Report No. 13, Natn. Inst. agric. Engng, Silsoe.

Sturrock F G, Cathie J, Payne T A 1977 Economies of scale in farm mechanisation. Occas. Papers No. 22, Economics Unit, Dept Land Economy, Cambridge Univ.

Tame P 1984 If it goes on the road it must conform. *Farmer's Weekly* Supplement – Farm Transport, Sept. 28, p 21.

Taylor A J 1984 Controlling costs. *Twenty-fourth Annual Conf. Report.* East Scotl. Coll. Agric., Edinburgh. pp 33–41.

Terratec 1982. *Ploughing performance predictor.* South Barnton Ave., Edinburgh.

Thornthwaite C W 1948 An approach toward a national classification of climate. *Geogr. Rev.* **38**: 55–94.

Turnage G W 1972 Tire selection and performance prediction for off-road wheeled vehicle operations. *Proc. 4th Int. Conf., Int. Soc. Terrain-Vehicle Systems, Stockholm.*, pp 61–82.

Tyldersley J B, Blackmore R A, Rumney R P 1974 Sequences of dry days for grass conservation in Lancashire and Yorkshire. Agric. Memo. No. 616, Meteorological Office.

Upton M 1964 Factors affecting turn time and rate of working of field machines. *J. agric. Engng Res.* **9**: 372–80.

Watson R D 1979 Choosing and using a respirator: dangerous dusts and vapours in agriculture. Bull. No. 18, North Scotl. Coll. Agric., Aberdeen.

Welford A T 1968 *Fundamentals of skill.* Methuen.

White D J 1975 Energy in agricultural systems. *Agric. Engr* **30**: 52–8.

Wilkins J D, Coleman R N 1971 High-speed field tractors, why? Paper No. 710686, Soc. Automotive Engrs, New York.

Williams D 1968 Introduction to the basic method. In Thornley G (ed.) *Critical path analysis in practice.* Tavistock, London.

Wilson R W 1971 Implement control by tractor driveline torque sensing. Paper No. 71-A609-NP, Am. Soc. agric. Engrs, St Joseph, Mich.

Wismer R D, Luth H J 1973 Off-road traction prediction for wheeled vehicles. *J. Terramechanics* **10**: 49–61.

Witney B D 1974 Tractor fuel economy. Adv. Leaflet No. 80, East Scotl. Coll. Agric., Edinburgh.

Witney B D, Beveridge J L 1975 Some economic aspects of ensilage mechanisation for beef production. *Agric. Engr* **30**: 12–17.

Witney B D, Morrison R R 1977 Of all the hay curing systems, barn-dried is best buy. *Power Farming*, June, pp 26, 37, 39, 41.

Witney B D 1981 Ploughs and ploughing: tractors and traction. Publ. No. 80, Scott. agric. Coll.

Witney B D, Eradat Oskoui K 1982 The basis of tractor power selection on arable farms. *J. agric. Engng Res.* **27**: 513–27.

Witney B D, Eradat Oskoui K, Speirs R B 1982 A simulation model for predicting soil moisture status. *Soil & Tillage Res.* **2**: 67–80.

Witney B D 1983 Mechanisation, the right power at the right cost. *Agric. Engr* **38**: 80–6.

Witney B D 1983 The conventional plough still has a lot going for it. *Arable Farming*, Nov., pp 34–5, 38.

Witney B D 1984 Selection of tractor power and machinery systems for the production of spring cereals. In Carr M K V (ed.) *Crop Establishment: Biological Requirements and Engineering Solutions. Aspects Appl. Biol.* **7**: 343–58.

Witney B D, Copland T A, Milne F 1984 Performance monitors for field work – economics and techniques. Paper presented at Ag Eng 84, Cambridge.

Witney B D, Elbanna E B 1985 Simulation of crop yield losses from untimely establishment. *Res. Dev. Agric.* **2**: 105–17.

Zegers D H A 1985 Ergonomic workplace assessment with the aid of a computer program. *Agric. Engr* **40**: 74–8.

Zoz F M 1972 Predicting tractor field performance. *Trans Am. Soc. agric. Engrs* **15**: 249–55.

Zoz F M 1974 Factors affecting the width and speed for least cost tillage. *Agric. Engr* **29**: 75–9.

INDEX